Certifiable $\mathcal{L}_1$ Adaptive Control for Helicopters

TECHNISCHE UNIVERSITÄT MÜNCHEN

LEHRSTUHL FÜR FLUGSYSTEMDYNAMIK

Certifiable $\mathcal{L}_1$ Adaptive Control for Helicopters

Magnus Bichlmeier

Vollständiger Abdruck der von der Fakultät für Maschinenwesen der Technischen Universität München zur Erlangung des akademischen Grades eines

Doktor-Ingenieurs

genehmigten Dissertation.

Vorsitzender: Univ.-Prof. Dr.-Ing. Manfred Hajek
1. Prüfer: Univ.-Prof. Dr.-Ing. Florian Holzapfel
2. Prüfer: Prof. Naira Hovakimyan, Ph.D.
                  University of Illinois at Urbana-Champaign (USA)

Die Dissertation wurde am 18. Mai 2015 bei der Technischen Universität München eingereicht und durch die Fakultät für Maschinenwesen am 09. Februar 2016 angenommen.

Bibliografische Information der Deutschen Nationalbibliothek

Die Deutsche Nationalbibliothek verzeichnet diese Publikation in der Deutschen Nationalbibliografie; detaillierte bibliografische Daten sind im Internet über http://dnb.d-nb.de abrufbar.

1. Aufl. - Göttingen: Cuvillier, 2016

Zugl.: (TU) München, Univ., Diss., 2016

© CUVILLIER VERLAG, Göttingen 2016
Nonnenstieg 8, 37075 Göttingen
Telefon: 0551-54724-0
Telefax: 0551-54724-21
www.cuvillier.de

1. Auflage, 2016
Gedruckt auf umweltfreundlichem, säurefreiem Papier aus nachhaltiger Forstwirtschaft

ISBN 978-3-7369-9281-8
eISBN 978-3-7369-8281-9

Zusammenfassung

Diese Arbeit beschäftigt sich mit verschiedenen $\mathcal{L}_1$-adaptiven Reglern für einen generischen Mehrzweckhubschrauber. Neben den Anforderungen an die Handling Qualities steht eine Strategie im Vordergrund, die eine zivile Zulassung des jeweiligen Regelungssystems erlauben könnte. Daher wird die geringst mögliche Komplexität verfolgt, die für Funktionalität und Flugleistung noch vertretbar ist.

Eine verkürzte Entwicklungszeit bei gleichzeitig verbesserter Überlebensfähigkeit des Hubschraubers bietet neue Chancen auf dem Markt. Die Überlebensfähigkeit in nicht-nominalen Zuständen bezieht sich hauptsächlich auf Situationen, in denen der Pilot mit erheblich verschlechterten Steuereigenschaften der Maschine konfrontiert ist. Gerade aber vor dem Hintergrund der Wirtschaftlichkeit neuer Systeme gilt besonderes Augenmerk in dieser Arbeit der technischen Ausführ-barkeit. Insbesondere die beschränkt verfügbare Rechenleistung in branchenüblichen Flugrechnern wird berücksichtigt.

Neben der adaptiven Augmentierung des Basisreglers werden in dieser Arbeit allein-lauffähige (standalone) $\mathcal{L}_1$-adaptive Regler beschrieben, ein Basisregler ist dabei nicht vorhanden. Über diese Unterscheidung von Augmentierung und standalone-Reglern hinaus kann in Zustands- und Ausgangsrückführung unterschieden werden. In dieser Arbeit werden Kombinationen dieser Kategorien gezeigt.

Die Implementierung wird ergänzt durch zahlreiche theoretische Überlegungen. So wird eine Modifikation des adaptiven Gesetzes, welches für die Zustandsrückführung entwickelt wurde, in die Ausgangsrückführung portiert. Diskutierte Szenarien beinhalten Sensornoise, Aktuatorsaturierungen sowie begrenzte Rechnerkapazität, speziell die zeitlichen Rechnerschrittweiten. Falls möglich sind dazu mathematische Beweise gezeigt, andernfalls beschreiben representative Simulationen die entscheidenden Effekte. Als Ergänzung und wichtiges Element der Zulassung werden Sensitivitätsanalysen von Designparametern auf Güte und Robustheit des Regelungssystems durchgeführt.

Die Arbeit schließt mit einer Zulassungsstrategie ab, die die theoretischen Beweise und die Leistungsfähigkeit des Regelungssystems berücksichtigt. Einige Überlegungen zur Robustheit, zum Beispiel gegen Abweichungen im Rechentakt, runden die Arbeit ab.

Zur Implementierung und Erprobung steht eine proprietäre, high-fidelity, gray-box Simulationsumgebung zur Verfügung.

Abstract

Various $\mathcal{L}_1$-adaptive controllers are designed for a generic utility helicopter. In addition to a required minimum level of handling qualities the most important prerequisite is the design of a system which could be certified for civil aerospace. Thus, a system with a minimal level of complexity to achieve adequate performance is sought.

Motivated by the demand and economic incentives of faster development cycles combined with enhanced survivability in off-nominal situations, emphasis in this thesis is put on the technical feasibility regarding industry standard in hardware and its performance limitations. Survivability refers to conditions, in which the pilot is challenged with significantly degraded handling qualities.

In addition to the adaptive augmentation of the baseline controller, standalone $\mathcal{L}_1$-adaptive controllers are shown. Aside from distinguishing between augmentation and standalone $\mathcal{L}_1$-adaptive control systems, another distinction is drawn between $\mathcal{L}_1$-adaptive controllers in state feedback and in output feedback. Combinations of these are addressed in this thesis.

The implementation with its practical considerations is supplemented by several theoretical considerations. This includes the porting of a modification of the raw piece-wise constant adaptive law (developed for state feedback) to output feedback. Furthermore, the presence of measurement noise, actuator saturations, and limited computational power are discussed – with rigorous proofs if applicable, in simulations otherwise. Supplemental to this and as a very important element of the certification strategy, a sensitivity analysis of design parameters on performance and robustness is conducted.

The thesis concludes with a certification strategy, that sums up rigorous proofs and the performance of the control systems in various conditions. Robustness considerations round off the thesis.

The control laws are implemented and evaluated in a high-fidelity proprietary gray-box simulation environment.

Preface and Acknowledgments

Acknowledgments

First and foremost, I want to thank Prof. Naira Hovakimyan. Furthermore, I am deeply grateful to Manuel Stölzle and Dr. Enric Xargay for their practical support. I would also like to thank Gunther Michalka for a seamless cooperation. Further thanks go to Prof. Chengyu Cao, Ronald Choe, Dr. Evgeny Kharisov, Dr. Zhiyuan Li, Dr. Hui Sun, Dr. Michael Kriegel, John Cooper, Christian Merkel, and Prof. Giulio Avanzini for their inputs. I would also like to express my gratitude to Steffen Greiser, Johannes Wartmann, Sven Lorenz, and Johann Dauer for insightful conversations, to Thomas from the police helicopter unit and my landlords Peter and Jerome for their marvelous support. My special thanks go to the Fedora-, FlightGear-, and related developers for providing free software.

The Demand for "Adaptive" Controllers

Faster development cycles with reduced time to market are becoming more important. Especially the iterative tuning with repetitive loop shaping and system identification in frequency domain is sought to be reduced. Tracking adaptively an explicit model allows for changing desired dynamics during flight for technical or handling quality evaluation. For online parameter identification with the purpose of online controller design, certain conditions have to prevail, e.g. persistent excitation. In general, a certain amount of information has to be present for identification. This scheme often collapses because of exactly these preconditions. Even without the requirement of parameter convergence, robustness issues occur (for which remedies exist), but despite deterministic adaptive laws, locally guaranteed performance bounds are a scarce occurrence among countless theories in adaptive control. While overcoming the slowness of statistical learning, it is important to maintain robustness, i.e. operating without high gain feedback to the control signal.

The Author's Comment on "Adaptive"

Historically, a controller was called adaptive whose feedback gain is adjusted with the control error. With the dilution of this idea in newer developments, discussions arose which controller can be called adaptive. Besides the hardly disputable fact of different levels of adaptation, the author recommends orientation at the original background – adaption in biology. There, adapting refers to life forms and structures that evolve over time and adjust to meet the needs of a changing environment or situation, be it in self organization or by active interpretation. In that sense, none of the current predictor based architectures is fully adaptive. The only adaptive part is the way how the hard-coded desired dynamics are achieved. A simple error integrator shows most of the characteristics of what currently is called adaptive. Yet any categorization driven too far is an impediment for advancing the state of the art.

magnus.bichlmeier@tum.de

Contents

List of Figures   viii

Acronyms   xi

Notation and Symbols   xii

1 Introduction   **1**
   1.1 Motivation 1
   1.2 Controller Requirements and Objectives 3
   1.3 Chapter Overview 5
   1.4 Contributions of this Thesis 6

2 Background   **8**
   2.1 Helicopter Dynamics 8
   2.2 System Description 12
   2.3 Offline System Identification 14
   2.4 The Baseline Controller 16
   2.5 Introduction to $\mathcal{L}_1$-Control 26
     2.5.1 The Idea 26
     2.5.2 Explanation to the Piece-Wise Constant Adaptive Law 32
     2.5.3 Introduction to Output Feedback for Non-SPR Desired Dynamics . 34
     2.5.4 Performance Effects of the Prediction Error: Issues and Solutions . 35
     2.5.5 Performance and Robustness 37
   2.6 Augmenting and Standalone $\mathcal{L}_1$-Controllers 38
   2.7 Internal Model Based Control 42

3 Design of the Input Channel to the Predictor   **45**
   3.1 Input Signal Merging 45
   3.2 Trim 47
   3.3 Decoupling of Cross-Couplings 47
   3.4 Structure 48
   3.5 Modeling of Response Lags 51

4 $\mathcal{L}_1$-Control for Pitch, Roll, Yaw   **53**
   4.1 In General 53
   4.2 State Feedback 54

4.3   Output Feedback … 56

4.4   Outer Loops for the Standalone $\mathcal{L}_1$-Controller … 58

5 $\mathcal{L}_1$-Control for Vertical Speed in Hover … **61**

5.1   Background … 61

5.2   Controller Core … 62

5.3   Command Signal Processing … 64

5.4   Extensions and Enhancements … 66

5.5   Alternative Structure … 67

6 Certification Strategy … **68**

6.1   Introduction … 68

6.2   Core Reasoning … 69

6.3   Additional Supportive Criterions … 71

7 Simulation Results … **74**

7.1   Frequency Domain: Baseline Controller … 76

7.2   Frequency Domain: Augmentation with State Feedback … 78

7.3   Frequency Domain: Standalone $\mathcal{L}_1$-Control in State Feedback … 80

     7.3.1   Raw Adaptive Law … 80

     7.3.2   Recursive Adaptive Law … 82

7.4   Frequency Domain: Standalone $\mathcal{L}_1$-Control in Output Feedback … 84

7.5   Time Domain: Standalone $\mathcal{L}_1$-Control in State Feedback – Raw Adaptive Law … 86

7.6   Vertical Speed Controller … 87

8 Conclusion and Future Work … **88**

8.1   Conclusion … 88

8.2   Future Work … 91

A Definitions … **92**

A.1   General Definitions … 92

A.2   Coordinate Frames … 96

B Alternatives for Decoupling of Cross-Couplings in Augmentation … **98**

C $\mathcal{L}_\infty$-Stability … **100**

C.1   Introduction and Conditions … 100

C.2   Proof of Theorem C.1.1 … 100

D Equivalent State Feedback Systems … **102**

E Performance Bounds in State Feedback … **104**

E.1   Definitions, Assumptions, Descriptions … 105

E.2   Reference System … 111

E.3   Core Proof … 112

    E.4 Proof of Lemmas 117
       E.4.1 Proof of Lemma E.2.1 117
       E.4.2 Proof of Lemma E.3.1 118
       E.4.3 Proof of Lemma E.3.2 120
       E.4.4 Proof of Lemma E.3.3 121
    E.5 Alternative, Recursive Adaptive Law 122

F Performance Bounds in Output Feedback for Non-SPR Desired Dynamics with a Recursive Adaptive Law   **126**
    F.1 Definitions, Assumptions, Descriptions 127
    F.2 Reference System 136
    F.3 Core Proof in Initial Broken-Loop Mode 136
    F.4 Core Proof in Standard Closed-Loop Mode 148
    F.5 Proof of Lemmas 160
       F.5.1 Proof of Lemma F.1.1 160
       F.5.2 Proof of Lemma F.1.2 165
       F.5.3 Proof of Lemma F.1.3 165
       F.5.4 Proof of Lemma F.2.1 165
       F.5.5 Proof of Lemma F.3.1 167
       F.5.6 Proof of Lemma F.3.2 168
       F.5.7 Proof of Lemma F.4.1 168
       F.5.8 Proof of Lemma F.4.2 and Lemma F.3.3 170
       F.5.9 Proof of Lemma F.4.3 172
       F.5.10 Proof of Lemma F.5.1 173
       F.5.11 Proof of Lemma F.5.2 175

G Signal Hedging with Saturation   **176**
    G.1 Introduction and Conditions 176
    G.2 Proof of Theorem G.1.1 177

H The Effect of a_{SP} on $|\tilde{x}|$, $|x_{ref} - x|$, $|x_{ref} - \hat{x}|$ **179**
    H.1 Description 179
    H.2 Example 180
    H.3 Explanation 182

I Controller Robustness against Time Step Variations   **187**
    I.1 Description 187
    I.2 Example 188
    I.3 Explanation 189

J Controller Robustness against Measurement Noise   **191**
    J.1 Description 191
    J.2 Example 193
    J.3 Proof of Theorem J.1.1 196
    J.4 Proof of Theorem J.1.2 197
    J.5 Proof of Theorem J.1.3 197

K Modeling of Inertia Effects with a Gyroscopic Term   **199**

L Alternative Predictor in an Error Space   **201**

M Alternative State Predictor Enclosing Baseline Controller States   **203**
    M.1 Definitions 203
    M.2 Core Statements 204
    M.3 Necessity of the Alternative Predictor 207
    M.4 Proof of Theorem M.3.1 207
    M.5 Example 209

N Examples of Practical Verification and Falsification   **212**

O Simulation Setup   **215**
    O.1 Concept 215
    O.2 Structure 216
    O.3 Parameter Scheduling in State Feedback 218
    O.4 Actuators 218
    O.5 Noise Modeling 220
    O.6 Miscellaneous 220

Bibliography   **222**

List of Figures

1.1 Photograph of a Mi-24 2

2.1 Helicopter drawing – conventional configuration 8
2.2 Qualitative natural step response of $\dot{h}(t)$ on a collective input in hover . . . 12
2.3 Exemplary helicopter eigenvalues 12
2.4 Exemplary frequency sweep 14
2.5 Baseline controller – pitch and roll 17
2.6 Proportional-derivative command module 17
2.7 Baseline controller – yaw 17
2.8 Heading hold controller 17
2.9 A typical notch filter 19
2.10 Anti-wind-up module 19
2.11 Exemplary sensitivity function 22
2.12 Exemplary loop transfer function (bode plot) 23
2.13 Exemplary Nichols diagram 23
2.14 Decoupling function 25
2.15 Concept of $\mathcal{L}_1$-control – simplified 26
2.16 Concept of $\mathcal{L}_1$-control – detailed 30
2.17 Elements of the trade-off: performance-robustness 38

3.1 Predictor input channel – signal *before* actuators 48
3.2 Predictor input channel – signal *after* actuators 48
3.3 Predictor input channel – variable saturation for signal before actuators . . 50
3.4 Predictor input channel – alternative trim structure 50
3.5 Predictor input channel – tip-path plane lag model 52

4.1 Implemented $\mathcal{L}_1$-controller 54
4.2 Broken-loop $\mathcal{L}_1$-controller 58
4.3 Outer loop in pitch and roll – attitude command 59
4.4 Outer loop in pitch and roll – rate command 59
4.5 Outer loop in yaw 59
4.6 Outer loop in pitch and roll – velocity command 60

5.1 Architecture of the vertical speed controller 64
5.2 Alternative architecture of the vertical speed controller 67

6.1 Effects of a_{SP} on the size of the adaptive law 72

7.1 Attitude tracking, baseline controller 76
7.2 Analytically computed sensitivity functions of the baseline controller . . . 77
7.3 Attitude tracking, augmented baseline controller 78
7.4 Sensitivity functions of the augmented baseline controller 79
7.5 Attitude tracking, standalone state feedback, raw adaptive law 80
7.6 Identification $d_{in}(t) \to u(t)$ in state feedback, raw adaptive law 81
7.7 Attitude tracking, standalone state feedback, recursive adaptive law 82
7.8 Identification $d_{in}(t) \to u(t)$ in state feedback, recursive adaptive law 83
7.9 Attitude tracking, standalone output feedback, recursive adaptive law . . . 84
7.10 Identification $d_{in}(t) \to u(t)$ in output feedback, recursive adaptive law . . . 85
7.11 Attitude tracking in time domain and actuator history – pitch 86
7.12 Attitude tracking in time domain and actuator history – roll 86
7.13 Rate tracking in time domain and actuator history – yaw 86
7.14 Vertical speed over time (integrated) for MTOM and MTOM/2 – *without*
    actuator position saturation 87
7.15 Vertical speed over time (integrated) for MTOM and MTOM/2 – *with* ac-
    tuator position saturation 87

A.1 Body-fixed coordinate frame 97

G.1 Signal hedging with saturation 177
G.2 Explanation to proof of Theorem G.1.1, Case 2 178

H.1 Effect of a_{SP} on $x(t)$ 180
H.2 Effect of a_{SP} on $\hat{x}(t)$ 181
H.3 Effect of a_{SP} on $\tilde{x}(t)$ 181
H.4 Effect of a_{SP} on robustness with additional delay 182
H.5 Effect of a_{SP} on terms in the error dynamics 183
H.6 Sensitivity of a_{SP} to $\tilde{x}(t)$ 184

I.1 Time step variation – CPU is slower 188
I.2 Time step variation – CPU is faster 189

J.1 Closed-loop $\mathcal{L}_1$-controller with noise 191
J.2 Transfer functions of noise in state feedback with raw adaptive law 194
J.3 Transfer functions of noise in state feedback with recursive adaptive law . . 195
J.4 Transfer functions of noise in output feedback 195

L.1 Error space controller 201

M.1 Trajectory $x(t)$ in augmentation, additional time delay of 0 ms 210
M.2 Trajectory $x(t)$ in augmentation, additional time delay of 10 ms 211
M.3 Trajectory $x(t)$ in standalone mode, additional time delay of 10 ms 211

N.1 Exemplary time histories of $x(t)$ and $\hat{x}(t)$ for pitch rate 213
N.2 Comparison of predictor dynamics to identified plant 213
N.3 Applying the baseline controller to predictor dynamics 214

O.1 Visualization screenshots 216
O.2 Simulation topology 217
O.3 Parameter scheduling: Interpolation and analytic approximations 219
O.4 Actuators 219

Acronyms

ACAH	Attitude Command Attitude Hold
ADS	Aeronautical Design Standard
AoA	Angle of Attack
BIBO	Bounded Input Bounded Output
CG	Center of Gravity
CMD	Command
CPU	Central Processing Unit
DOF	Degree of Freedom
DVE	Degraded Visual Environment
IMU	Inertial Measurement Unit
LTI	Linear Time Invariant
LTV	Linear Time Variant
MIMO	Mulitple Input Multiple Output
MRAC	Model Reference Adaptive Control
MTE	Mission Task Element
MTOM	Maximum Takeoff Mass
NED	North East Down
PIO	Pilot Induced Oscillations
RCAH	Rate Command Attitude Hold
RMS	Root Mean Square (Standard deviation with mean=0)
RPM	Revolution Per Minute
SISO	Single Input Single Output
SPR	Strictly Positive Real
UAV	Unmanned Aerial Vehicle

Notation and Symbols

w	Column vector of w_i; $i = [1, n]$
w^T	Row vector of w_i; $i = [1, n]$
$\hat{w}$	Predicted (computed or estimated)[1] value of vector w
$\tilde{w}$	Error between predicted and real value of vector w
$h(s)$	Polynomial in Laplace domain
$n_{Poles}(H)$	Number of poles of a transfer function $H(s)$
$n_{Zeros}(H)$	Number of zeros of a transfer function $H(s)$
$\wp(h)$	Order of $h(s)$
$reldeg(H)$	Relative degree of a transfer function $H(s)$, $n_{Poles}(H) - n_{Zeros}(H)$
$H_n(s)$	Numerator of a transfer function $H(s)$
$H_d(s)$	Denominator of a transfer function $H(s)$
$tr(A)$	Trace of matrix A
$\lambda_{max}(A)$	Biggest eigenvalue of matrix A (same with min)
$\|w\|_2$	2-norm of vector w
$\|w\|_\infty$	∞-norm of vector w
$\|w_\tau\|_{\mathcal{L}_\infty}$	Truncated $\mathcal{L}_\infty$-norm of $w(t)$
$\|w\|_{\mathcal{L}_\infty}$	Short notation for $\|w(t)\|_{\mathcal{L}_\infty}$
iff	Shortened for "if and only if"
$*$	Convolution operator.

---

[1] In some cases $\hat{w}$ is a variation of the prediction for including other variables.

ω^{IB}	Angular velocity ω of the frame B with respect to the frame I
$(\omega^{IB})_B$	Angular velocity ω of the frame B with respect to the frame I denoted in the frame B
$(\dot{\omega}^{IB})_B^B$	Derivative with respect to frame B of angular velocity ω of the frame B with respect to the frame I denoted in the frame B

$\mathbb{I}$	Identity matrix
$\mathbb{R}$	Set of real numbers
$\equiv$	Identity symbol
$\square$	End of proof

Vectors and matrices are denoted as any other variable. If the content requires explanations, the dimensions are given.

In this thesis, "$\mathcal{L}_1$-adaptive Control" is shortened to "$\mathcal{L}_1$-control". This does in no way relate to $\mathcal{L}^1$-control from robust control theory.

The term "fly-by-wire architecture" represents also "fly-by-light" architectures.

Several ideas of this thesis with preliminary results are first published in [1] and in [2] by the author. The baseline controller is the subject matter of the author's diploma thesis [3].

Drawings or photographs depicting a specific type of helicopter are for conceptual purposes only and do not represent any real or simulated type.

Chapter 1

Introduction

A generic utility helicopter of conventional configuration with main and tail rotor is considered. Several realizations of $\mathcal{L}_1$-control for attitude and vertical speed control are shown. A high-fidelity simulation in place of a real helicopter serves as research platform, treated as gray-box simulation. Publications of preliminary results are: [1], [2], and in particular the baseline controller: [3] (the author's diploma thesis).

1.1   Motivation

If the plant dynamics and their changes are not sufficiently well known, adaptive elements may be desirable. A controller is aimed at that contains explicitly the predefined desired dynamics, wherein tracking is done adaptively. In addition, the desired dynamics need to be provided with mechanisms to ensure feasibility, that is to account for time delays, input saturations, and the limited input channel bandwidth – present in any physical system. The combination of feasible desired dynamics with an adaptive tracking strategy holds out the prospect of two major benefits:

1. Maintaining handling qualities in adverse conditions and thus enhancing survivability;

2. Reducing the development effort;

Survivability of the helicopter in this context reduces most notably to maintaining handling qualities. It may be questionable whether keeping a helicopter aloft in case of severe damage is significantly more likely with a different controller structure. It is of particular interest however to *maintain handling qualities* in DVEs (degraded visual environments), MTEs (mission task elements) with divided attention e.g. delivering or picking up loads, or combat situations – in general operating near ground. For avoiding obstacles, predictable behavior of the vehicle is crucial, especially in military operations, where flying very low and fast is a frequently applied tactic to escape hostile fire. If degraded performance cannot be avoided, pursuing dynamics scaled to a lower bandwidth may be the best option. This means keeping the same behavior (e.g. linear, first order) but with different velocity (i.e. gain or bandwidth).

Furthermore, in some situations loss of control can be prevented only in a very short time window. A fast acting controller as well as retaining handling qualities for a safe recovery are crucial, especially in the inner loop. A vehicle suddenly and without clear warning degrading from Level 1 to Level 3 on the Cooper Harper Rating Scale is believed to be worse than an aircraft being Level 3 from the beginning.

Helicopter dynamics are complex even without any failure and finding a linear design point may be elusive, e.g. for the asymmetric, weak and time varying directional stability. Another example of dynamics, that are hard to capture and hardly quantifiable, are cross-couplings. For these cases, an adaptive model following strategy can be helpful and simplify the design procedure to meet the increasing requirements of recent safety specifications.

FIGURE 1.1: An Mi-24 (Photo from the author's collection)

This tempts to aim at a care-free handling approach as it is practiced with unstable fighter jets, this however can be accompanied by an unreasonable high effort – if possible at all – for helicopters. Automating the prediction of the vortex ring state or the pitch-up phenomenon reliably for example can be tedious up to impossible, leaving it to pilot training to avoid it. Given the fact that escaping the flight envelope can normally be averted but the risk is impossible to be eliminated, recovery from these conditions back into the normal operational flight envelope is a desirable capability of the controller. When encountering enemy fire, no pilot can be expected to stick to low-frequency inputs or safe flight strategies.

The second point has obvious benefits also beyond manned civil planes. Cutting costs in the controller development may be attractive for the low-cost UAV industry, for instance.

An adaptive model following controller may be utilized as cross-platform controller for many types of helicopters. Then, tuning reduces to adjusting the implemented model of the input channel as well as the desired closed-loop performance specified in the predictor. This may be realized by a conservative choice of the desired performance, expected to be within robustness margins for an entire class of helicopters. However, the new methodology does not relieve of the required understanding of the *helicopter's* peculiarities and performance bounds.

1.2   Controller Requirements and Objectives

From modern adaptive controllers, fast adaptation, reasonable design effort, and sufficient time delay margin representing robustness are expected. Performance, in particular in the transient, is required to be deterministic and locally guaranteed. The controller is applicable to analysis of safety-critical systems, a fact that imposes strict conditions on functionality and reliability.

The Choice of Linear Desired Dynamics

Linear low order systems are considered easy to operate and predict for a human being. Especially the first order system represents well predictable dynamics without any overshoot. Extrapolation of nonlinear relations can instead be a very difficult task for humans. As a result, the desired dynamics are chosen to be linear and with an order as low as possible. Moreover, theory for linear desired dynamics is slightly simpler. Parameter scheduling can be applied for adjusting the desired dynamics. The plant dynamics are allowed to be nonlinear in all cases as it is shown in subsequent chapters.

Handling Quality Requirements

A landmark for controller optimization are the specifications stated in ADS-33 ([4]), covering important conditions for satisfying handling qualities. In general, a fast response is desired, without large overshoots and with as little time delay as possible. The controller with the fastest response however is not necessarily the best one for pilots. A predictable and reasonable fast response is preferred to an overly aggressive one. Especially an aggressive attitude disturbance rejection would lead to discomfort when hitting gusts frequently. The riding qualities as described in [5] would suffer in these cases. This issue refers e.g. to outer loops of a (rate) inner loop adaptive part or the baseline controller in case of augmentation. Additionally, large phase lags and an overly high stick sensitivity are significantly hurting handling qualities, where phase lag is a significant driver of PIOs (pilot induced oscillations).

Systemic Controller Requirements

General requirements are applicable to controllers to be certified:

1. High frequencies (compared to actuator bandwidth) in the control signal, that have virtually no effect on the plant output, are to be avoided for saving actuator wear, despite the fact that cyclic control usually needs very little energy [5].

2. Measurement noise must not lead to instability nor be significantly propagated through to the control signal.

3. Input saturation must not lead to instability or overly high performance loss apart from the missing control authority (cf. anti-wind-up architectures for integrators in PI-controllers).

4. The software may be implemented on contemporary hardware certified for aerospace. This entails limits on complexity (lines of code) as well as on numerical precision and the largest number possible to be processed. Besides, the code should be executed by a discrete solver with guaranteed numerical stiffness and precision properties.

5. A deterministic and repeatable nature of the algorithm. As opposed to unpredictable offline solvers, the controller is supposed to finish repeatedly every elementary task after $\Delta t = T$.

6. Performance guarantees during the transient response.

7. A verifiable robustness metric.

8. It should be possible to demonstrate the meaningfulness of the algorithm by formal mathematical methods.

9. Robustness against or active inclusion of input time delays.

For a certification approach, requirements for software in DO-178 (version 'C' at the time of writing) apply.

An explicit failure detection is sought to be avoided.

1.3   Chapter Overview

The motivation in Chapter 1 is followed by a list of requirements, shown in section 1.2. The compliance with the requirements is discussed in Chapter 6.

Chapter 2 provides background information about helicopter dynamics, the baseline controller, its design with the help of system identification, and an introduction to $\mathcal{L}_1$-control. Among many alternatives, one selected combination of solutions, the *primary architecture*, is presented in the main part, Chapters 3 to 5. It is evaluated in Chapter 7, "Simulation Results".

Chapter 3 describes modules for the input channel design, covering elements for saturation and signal hedging.

Chapters 4 and 5 show the realization of $\mathcal{L}_1$-control for the controller in pitch, roll, yaw and for vertical speed, respectively.

Chapter 6 serves as set of recommendations how to tackle certification for civil aerospace.

Chapter 8 sums up the most important results and provides an outline of additional efforts that can be undertaken.

Appendix A introduces definitions and some terminology.

Appendix B refers to alternative structures of the primary method of handling decoupling of cross-couplings.

Appendix C provides a mathematical background to $\mathcal{L}_\infty$-stability.

Appendix D shows various but equivalent forms of state predictors.

Appendices E and F show the general performance and stability proofs by formal mathematics.

Appendix G provides a formal theorem for the validity of signal hedging in the predictor input channel.

Appendices H, I, J are dedicated to a robustness and sensitivity evaluation of design parameters.

Appendices K, L, M show alternative structures not being included in the primary architecture.

Appendix N shows examples of system verification.

Appendix O explains the simulation setup.

The appendix is to be understood not only as background information, but as subject matter, which, if it had been presented in the main part, would have confused the reader. However, it includes important – if not the most important – information.

1.4 Contributions of this Thesis

The thesis tries to be complete in all relevant aspects of introducing adaptive control in civil aerospace. Thus, handling qualities, signal characteristics, implementation, structural interaction, sensor noise, input channel saturations, and rigorous formal mathematics explaining the meaningfulness of the algorithms are addressed.

A modified piece-wise constant adaptive law is ported to output feedback, allowing for output feedback to perform similar to state feedback in the simulations without higher sampling rates.

Existing formal proofs of theoretical performance bounds are modified: A recursive adaptive law is included in output feedback, an initialization procedure is merged into the proof of the performance bounds of output feedback, a different strategy in the proof of the performance bounds of state feedback is shown, and some simplifications are achieved. Sensor noise is included in the formal proofs with the help of separate noise transfer functions.

The trade-off between performance and robustness is specified. It is shown that in $\mathcal{L}_1$-control the shaping of the error dynamics and the amount of modeled time delay are important elements in the trade-off, whereas the choice of the filtering structure bandwidth is largely fixed by actuators and closed-loop system bandwidth. Furthermore, it is shown that for the piece-wise constant adaptive law in scalar systems, slow error dynamics are better performing and less robust. Guidelines for the choice of the filtering structure bandwidth are presented.

The propagation of the prediction error (caused by undesired dynamics and external disturbances) to the tracking error with the role of augmentation is addressed. In this context, a new understanding of augmentation as exclusively aiding the adaptive controller in preventing disturbance propagation to the tracking error via the prediction error is suggested.

The model following nature of the $\mathcal{L}_1$-controller in comparison to the baseline controller and the robustness implications thereof are considered, while referring to adaptive and nonadaptive properties.

A simplified and an extended predictor (including the baseline controller states) are compared.

Systematic input channel design guidelines are presented. Rigorous conditions are provided for hedging signals contained in the total command vector of the input channel signal.

A special structure for a vertical speed controller is proposed. A seamless activation, robustness against mass changes, hedging of trim inputs while keeping the software implementation effort low are the most important features.

An architecture controlling the error between the desired and the real dynamics is shown. A number of minor findings, mostly summed up in Chapter 8.1.

Chapter 2

Background

This chapter provides a broad outline of helicopter dynamics, the baseline controller, system identification, the concept of $\mathcal{L}_1$-control, and some other basic insights.

2.1 Helicopter Dynamics

This section introduces the reader to fundamentals of helicopter flight dynamics. A comprehensive description can be found in e.g. [5], [6], or [7] (in German language). The statements herein primarily refer to a main-tail rotor configuration depicted in Figure 2.1 (rotor blades in some trim position), but are mostly applicable to other configurations as well.

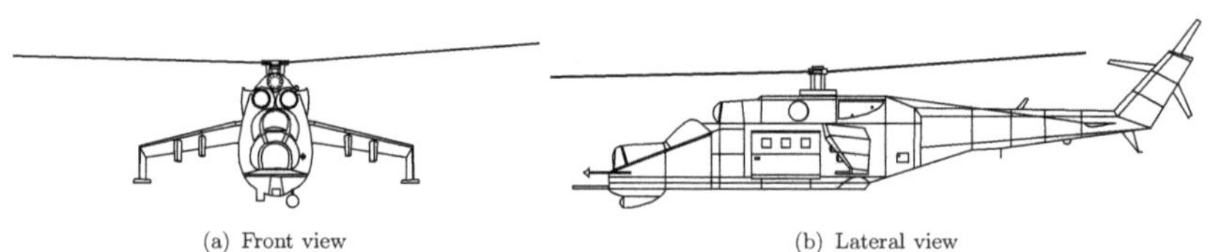

(a) Front view　　　　　　　　(b) Lateral view

FIGURE 2.1: Helicopter drawing – conventional configuration

Speaking of developing a helicopter is to a great extent equal to speaking of developing the rotor. The rotor is a system of rotor blades, spinning with approximately one constant RPM (revolutions per minute) or deliberately slightly varying RPM in modern types. The advancing blade encounters higher aerodynamic velocity in forward flight than the retreating one. To equalize the lift in forward flight, articulated[1] blades are used, which differs from a propeller. The increase of lift in the advancing blade is compensated for by flapping, i.e. a blade movement perpendicular to the rotor plane.

A flapping hinge requires an additional lagging hinge for allowing the DOF (degree of freedom) in the rotor plane at the blade root for avoiding large moments due to flapping. Accompanied with flapping is the radius reduction of the blade CG (center of gravity), implying a velocity change due to momentum conservation. Together with the third motion, the feathering motion which is the immediate control mechanism for helicopter rotors, every blade has three degrees of freedom: flap, lag, feather. Hence, a rotor blade is a pendulum under the dominant influence of centrifugal force (gravity is small compared to centrifugal force). This structurally flexible "pendulum" experiences forces (drag, lift, hinge moments, ...) and damping (aerodynamic, structural, artificially incorporated in hinge-dampers, often in lag motion due to the smaller aerodynamic damping compared to flap). The swash plate is the element that translates commands from the non-rotating airframe to the rotating rotor. The collective input changes the AoA (angle of attack) collectively, i.e. all blades by an equal amount. The cyclic input implies with every revolution a periodic change to the AoA. With the analogy to the pendulum, a phase lag occurs from a changed AoA to the peak of the succeeding flapping motion. If there is only one central hinge (as seen in two-bladed helicopters), this phase lag is 90 *deg* for see-saw rotors and less for a hinge offset > 0 and hingeless rotors. A periodic flapping motion, where the period coincides with rotor RPM tilts the rotor plane and with it, the thrust vector. Tilting the thrust vector out of the CG, a moment is generated that tilts the airframe and with it, the rotor plane. This new thrust vector has a horizontal component (additional to trim) that causes the helicopter to accelerate in the horizontal plane. The loss in the vertical thrust component can be compensated for by a higher collective input. This is the primary mode of control, the initial change in AoA however has significant effects, too, which are most evident in hover. See also [8].

Tilting the rotor plane has effects on the fuselage and vice versa. Fuselage and rotor disc can oscillate against each other. This is more visible in roll due to the significantly lower inertia than in pitch.

With the excitation of the blade flapping mode (similar to some force on a pendulum in gravity) the signal propagation from the actuators to the actual moment on the airframe is highly dynamical. Moreover, the rotor acts as a frequency filter.

For increasing responsiveness, hingeless (sometimes even bearingless) rotors are built. The hinges are replaced with flexible structural elements which can translate moments. This has

---

[1]... or semi-rigid or hingeless with an elastic DOF (degree of freedom), all however with limited capability of transferring moments from the blade to the hub.

several effects: A much more responsive rotor with better (relatively stronger[2]) moment generation stands against harming stability in forward flight. A hingeless rotor is more prone to the pitch-up phenomenon.

The pitch-up phenomenon has a number of contributors: In high speed forward flight, the advancing blade encounters higher lift than the retreating one, meaning that the rotor tends to flap upwards up to 90 *deg* later, i.e. at the front of the helicopter. The more this unwanted tilt of the rotor plane and with it the thrust vector causes a nose-up moment, the higher the AoA of the advancing blade, amplifying the effect and therefore destabilizing the helicopter. Increasing rotor thrust in the collective channel causes more downwash from the rotor to hit the horizontal stabilizer and thus reducing its stabilizing effect.

The varying effects of rotor downwash to the vertical and horizontal stabilizer apply to the tail rotor as well. It serves the purpose of yaw control besides compensation for torque of the main rotor and engine. Hence, any change in the collective input is a disturbance for yaw control, alleviated by feedforward elements that increase tail rotor thrust with main rotor thrust. Being exposed to the rotor downwash and fuselage wake, the tail experiences strong disturbances due to varying flow directions and phenomenons like tail shake can be excited. Also for the tail rotor a vortex ring state exists, where vibrations, marginal controllability and loss of thrust are the consequences.

The facts mentioned so far indicate that a rotor cannot be described as a gyroscope since besides the blades' degrees of freedom flap, lag, and feathering, the rotor blades are flexible and are bent significantly. Many effects however can be observed similar to the gyroscope simply by the fact of a fast rotating mass[3]. Regressing (adverse to the rotational direction) flap or lag modes appear as nutation and precession.

These gyroscopic effects imply strong couplings on the rotor system. Other sources of cross-couplings are the above mentioned lift difference for exciting the flapping motion (advancing the flapping effect for usually 60..90 *deg*), that is the phase lag that cannot be fully compensated for by design as it is varying over flight conditions, the aerodynamic couplings of e.g. the tail, and many more effects mostly of the rotor. The swash plate is integrated only with the expected offset.

The trim attitude for a helicopter is determined by a number of influences. Some of these are the CG position, the aerodynamic velocity vector, and design traits like the vertical position of the tail rotor. With the tail rotor generating thrust in the horizontal plane,

---

[2]Helicopters with a see-saw rotor (central flapping hinge) often have a wide airframe as with missing moments the payload is confined to a small area in the longitudinal direction.

[3]Rotor blades are not designed to be as light as possible, but for controllability reasons heavier and for much inertia for a safer transition into autorotation.

the main rotor compensates for it. Hence, a helicopter with a CG in the geometric lateral center lands always with one side of the skids first. Besides roll, the pitch trim attitude traverses strong changes over CG position and airspeed – in hover it shows a strong nose-up attitude.

In most flight regimes, the air flow is oriented downwards through the rotor, where the rotor causes a pressure jump from the upper to the lower side of the rotor while air flow velocity remains constant. For fast descents and in case of engine failure such that thrust cannot maintain altitude any longer, the helicopter can transit into autorotation, where the flow direction is reversed, so that the air flow through the rotor keeps the rotor spinning for sufficient thrust to limit the vehicle sink rate. This state is called the wind mill brake state. The sinkrate during steady state autorotation is stable as increased drag with higher RPM slows down the rotor and less RPM increase the sink rate which again accelerates the rotor. Especially with hingeless rotors, the helicopter is still well controllable, however with changed dynamics e.g. damping and input gain.

The air flow through the rotor is sought to display a clear direction. If in slow or zero forward speed the aerodynamic velocity caused by the sink rate is close to the induced velocity of the rotor, the helicopter has entered the vortex ring state, where chaotic flows enter and leave the rotor in both directions. This state is to be avoided by the pilot as huge sink rates build up quickly and controllability suffers severely.

In conclusion, the helicopter is a very complex dynamical system due to its rotor dynamics. Multiple effects add to significant vibration levels (much higher than in most fixed-wing planes), strong interactions of aerodynamics with structure and by the lack of predictability (non-steady state aerodynamics) to an inevitable amount of unmodeled dynamics in simulations (or later predictors).

Aerodynamically, the motion of the rotor blades causes very diverse behavior over vehicle forward speed and even over one revolution. Blades work in a large range of Reynolds and Mach numbers. Local stall in the retreating blade, transonic flow in the advancing blade (e.g. buffeting)[4], high angles of attack, yawed flow, blade vortex interaction (by the blade approaching next), rotor wake interaction, and blade fuselage interaction are only a few effects to be mentioned.

In addition, structural modes become a serious issue due to the broad and intense excitation of the rotor. Controllers are equipped with notch filters to avoid excitation, especially feedback in the critical frequencies.

The heave motion in hover is controlled by the collective lever.

---

[4]Blade tip velocities are in hover typically at about 0.6...0.66 Ma.

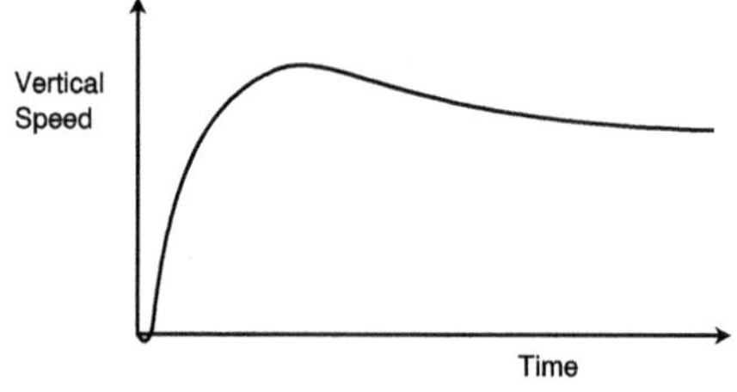

FIGURE 2.2: Qualitative natural step response of $\dot{h}(t)$ on a collective input in hover

Figure 2.2 shows a slightly simplified step response to a collective input in hover. The overshoot is explained by a higher airflow building up through the rotor which after some time decreases thrust. Naturally the response is of higher order than first or second order. The non-minimum phase characteristic undershoot arises in the time span when the new coning angle builds up.

2.2   System Description

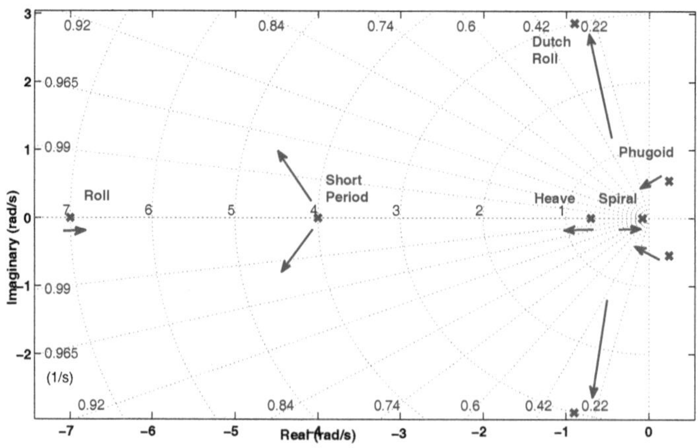

FIGURE 2.3: Exemplary helicopter eigenvalues in forward speed with their tendencies
with increasing forward speed

Quantification of Helicopter Dynamics

Figure 2.3 plots open-loop eigenvalues of a fictive example helicopter, flying with moderate forward speed. The arrows point in a possible direction of pole movement when increasing forward speed of the vehicle. The eigenvalues, direction and intensity of their variation are strongly dependent on the respective helicopter design, but first and foremost on the configuration and size. The rotor design as well as position and size of the horizontal tail plane are very important factors for instance. Examples for the variation of eigenvalues over speed for the BO-105, Lynx and Puma are given in [5]. In hover for instance, the location of the eigenvalues looks very different to what is shown in Figure 2.3.

Overview of Modeling Techniques

Depending on the purpose, a helicopter can be modeled in many ways. For endurance analysis, rotational dynamics do not play an immediate role – the helicopter as point mass is sufficient. In case rotational dynamics are considered, the simplest model is a rigid body 6-DOF model. This implies the neglect of rotor dynamics. For steady state power analysis in hover and vertical flight, conservation laws combined with Bernoulli equations for fluid flow can be applied. For a more detailed analysis, blade element theory can be utilized for incorporating the rotor dynamics. Finally, rotor-fuselage interactions, dynamics of other subsystems (e.g. actuators), non-steady state aerodynamics, structural modes, and their combination to aeroelastics can be considered.

For analyzing existing dynamic objects, system identification suits. Linear state space models and equivalently transfer functions can be identified. After fixing the structure of the model by physical insight, parameters are approximated by parameter identification. Despite nonlinear dynamics, "most" of the response can be captured by low order linear models. With "most" being a weak description of accuracy, a control goal of rendering the rate response first order is considered a reasonable and feasible goal. Any response however can at best be expected with some time delay and with small non-minimum phase effects.

If a linear state space model of a helicopter for one flight condition is to be obtained, decoupling longitudinal and lateral motions provides in general unsatisfactory accuracy. The couplings are strong enough to influence other axes significantly.

Further Considerations

If only dynamics between actuator and body-fixed angular rates are considered, there is no need to adapt for unmatched uncertainties.

Apart from couplings, the inputs are not redundant, i.e. the four inputs are mapped to the four-dimensional input vector space. In good approximation, the system can be modeled input affine, i.e. the input $u(t)$ enters the system linearly. A helicopter in the conventional main- tail rotor configuration is a non-holonomic system, but controllable. Accelerations in the horizontal plane can only be achieved by attitude changes away from the trim attitude (apart from the tail rotor that induces a force in the horizontal plane).

2.3 Offline System Identification

System identification is used herein only as a tool without the ambition to modify or improve. Hence it is described marginally here, backgrounds can be found e.g. in [9].

The basic steps for offline identification are excitation of the dynamics, recording the data of input and output signals for a later offline analysis, signal processing, and analysis. A form of the Fourier transformation is used to transform the data sampled in time domain into frequency domain, e.g. the Chirp-Z transformation. Windowing techniques are applied to obtain a frequency response. If necessary, the frequency response can be used for fitting a parametric model, i.e. transfer functions or state space models. Both, a frequency response and transfer functions (or state space models) imply linear behavior and therefore have limited but often sufficient accuracy. Verification of the models can take place in time or frequency domain.

This methodology of identifying in frequency domain and the contingent fitting of parametric models is the technique used most at the time of writing.

To excite all relevant modes, a frequency sweep with exponentially increasing frequencies is applied to either the commanded angular rate or the commanded attitude. Figure 2.4 shows a typical frequency sweep. If applied to rate commands, a short signal in the rate command (as shown in Fig. 2.4) for a small attitude change is added to the initiated sweep in order to shift the response to be around a trim attitude. The integrated sine wave of the rate would otherwise result in the attitude to be exclusively above or below the initial trim attitude. For a constant amplitude of the attitude, the amplitude of the rate is growing with the frequency as factor.

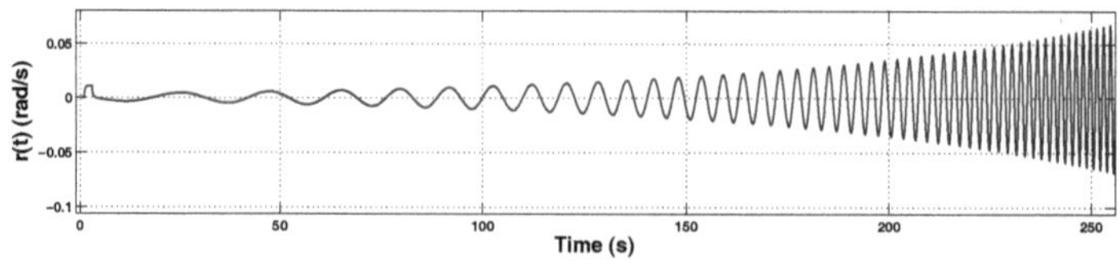

FIGURE 2.4: Exemplary frequency sweep applied to the rate command

Identification of rotorcrafts is challenging due to the high vibration level, highly coupled dynamics, unstable modes, etc..

It is recommended to adhere to the following:

- Data collection with a deactivated (baseline) controller is desirable to avoid correlations. A weakened controller is also often possible; weak gains however may have adverse effects due to rotor dynamics, the lead-lag eigenmode in particular.

- Identified correlations due to couplings can be mitigated by applying MIMO-identification techniques.

- If on the axis to be identified a computer generated frequency sweep (sinusoidal signal with exponentially increasing frequency) is applied to the input, a good method to eschew correlations from the other axes is to stabilize it manually with uncorrelated inputs. The frequency content for stabilizing off-axes should be decoupled from the on-axis.

- If achievable closed-loop desired dynamics are to be identified, the baseline controller is active and defines the assumingly decoupled closed-loop dynamics. SISO identification in both cases is justified.

- Despite the nonlinear behavior of helicopters (it may be approximated linearly very well), only linear identification is applied. Identification of nonlinearities can easily introduce more distortion than improvements due to its complexity.

- To capture basic dynamics, transfer functions of low order (e.g. 2...5) can be sufficient for e.g. implementing the identified transfer function as predictor dynamics ($=$ desired dynamics in $\mathcal{L}_1$-control) for augmentation. The lowest acceptable order dependent on the identified axis and on the helicopter type is aimed at.

- Contingent on the required accuracy of the identification, it is conducted at several points of the flight envelope. Most often, indicated airspeed is chosen as parameter to vary.

- When fitted to a frequency response, the structure of the parametric model is to be physically reasonable. It is desirable that also parameters obtained by the optimizer are physically meaningful, e.g. the roll time constant is usually approximately known and is expected to capture most of the roll dynamics.

2.4   The Baseline Controller

This section shows the baseline controller, as it is underlying in augmentation. However, it contains certain elements that are reused for standalone $\mathcal{L}_1$-controllers, e.g. the processing of trim signals or turn coordination. Key ideas, equations, and similar figures of this chapter are also presented in [1] and [3].

Architecture

The baseline controller is an enhanced PID-controller for pitch, roll, and yaw. The concept shown is relevant for fly-by-wire architectures but small modifications facilitate the application on other systems.

Separate gains for feedforward and feedback control are applied except for the error integrator gain. The term "PID" is slightly abused as the differential term "D" is not a real derivative of the attitudes "P" (proportional feedback), but measured body-fixed angular rates. The integral term "I" integrates attitude errors.

The **yaw controller** presents a slightly different architecture compared to pitch and roll. The reason is the different handling of the attitude, namely the heading. The terms "P","I","D" in yaw control do not reflect the classic nature of a PID-controller; rather the rate is controlled by a PI-controller, using only rate sensor information. This entails a different behavior compared to proportional feedback in rate and attitude. In case of an impulse-like disturbance, a PI controlled rate will go to zero but less likely go back exactly to the original attitude after the disturbance than in case of attitude feedback. The actual integral part is the integral of the heading error. Notice from Figure 2.8 that the command coincides in one mode with the actual measurement, meaning that this controller part adds zero in this case. When switching the heading hold mode on or off, the integrator is initialized. Heading hold is switched off in case of a pedal input by the pilot, if turn coordination is active, or during aerobatics mode, where attitude control is switched off. Then, the heading information is discarded.

The block "Trigonometry" in Figure 2.5 can be reduced to $1/cos(\Phi)$ for the pitch axis and is left blank in roll. This represents to some extent the coordinate system transformation from the Euler angles to the body-fixed system, as the commanded attitude refers to the latter system. Commanding Euler angles is counter-intuitive for a human pilot.

Figure 2.6 describes a possible command module for Figure 2.5, where $ß = 2\zeta w$ ('ζ' is the relative damping, 'w' the frequency). In yaw, the pedals connect to rate command only. If an additional command filter is applied, in general a second order filter for attitude commands and a first order filter for rate commands is proposed.

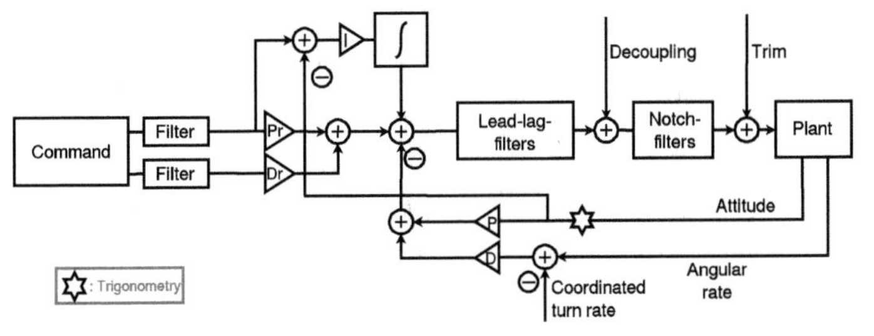

FIGURE 2.5: Baseline controller structure for pitch and roll

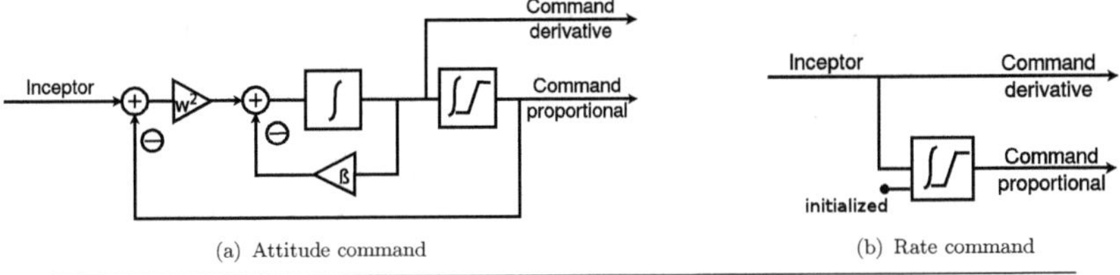

(a) Attitude command (b) Rate command

FIGURE 2.6: Proportional-derivative command module

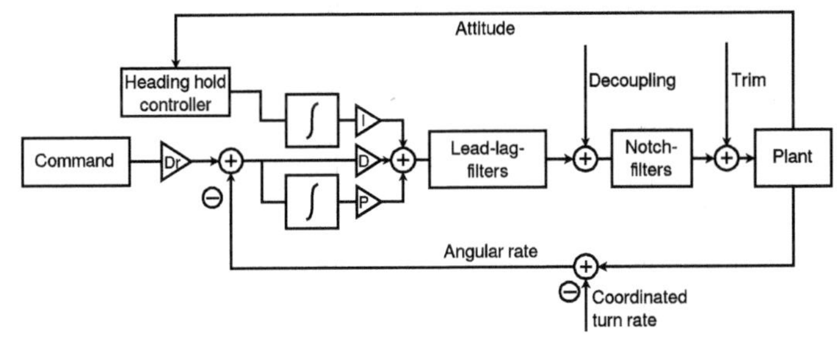

FIGURE 2.7: Baseline controller structure for yaw

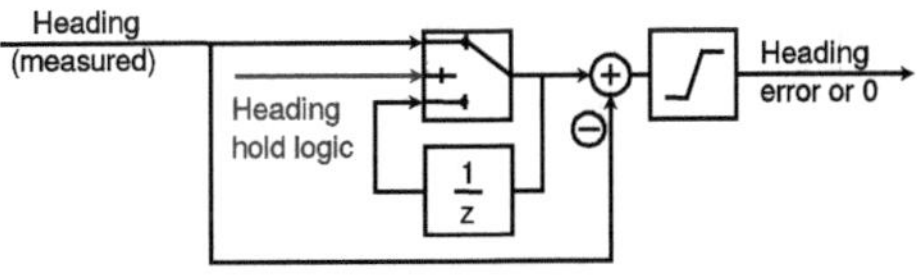

FIGURE 2.8: Heading hold controller

Reentering normal flight conditions (with respect to small Euler angles), feedforward filters and the attitude error integrator are reinitialized in all channels. Note that in Figure 2.5 two separate filters are implemented for attitude and rate feedforward control. This is due to the reinitialization of the filter in the attitude channel.

In addition to the PID structure, multiple second order lead-lag filters are added as a means of loop shaping. They have the following form that allows for being either a lead or a lag compensator:

$$LL(s) = \frac{(1/\omega_n^2)s^2 + 2\zeta_n/\omega_n + 1}{(1/\omega_d^2)s^2 + 2\zeta_d/\omega_d + 1} \tag{2.1}$$

with $0 < \zeta < 1$ being the relative damping and ω some frequency.

As mentioned in Chapter 2.1, a helicopter exhibits strong cross-couplings between different axes. This relates to command couplings and rate couplings. To attenuate cross-couplings, decoupling signals are added.
At this point, all signals with frequency content relevant to excite structural modes are added.

One or more notch filters are then added against excitation of resonances. One of the most important is the lead-lag eigenmode of the main rotor. A notch filter may be implemented by:

$$N(s) = \frac{s^2 + (\zeta_q/q)s + \omega_q^2}{s^2 + (1/q)s + \omega_q^2} \tag{2.2}$$

where $0 < \zeta_q < 1$ is the relative damping to set the amplitude, q is called the q-factor and defines the width of the passband, ω_q is the center frequency. Figure 2.9 shows a bode diagram of a typical notch filter.

Finally, trim signals are added with low frequency content and are thus irrelevant to decoupling and notch modules. There are two kinds of trim: 1) Actuator position trim; 2) Attitude trim;

This trim as shown in Figure 2.5 and Figure 2.7 is directly part of the total command vector of the actuator position. It acts as the "deliberate input disturbance" in the sense of a feedforward signal. It causes the controller signal to be "around zero".
The second trim signal is part of the attitude command to seek acceleration-free flight (apart from commanded accelerations) for alleviating pilot workload in ACAH mode (attitude command attitude hold). It maintains a nose up attitude and a bank angle that compensates for the approximately horizontal tail rotor thrust in hover, for instance. Both trim signals are varying with different flight conditions.

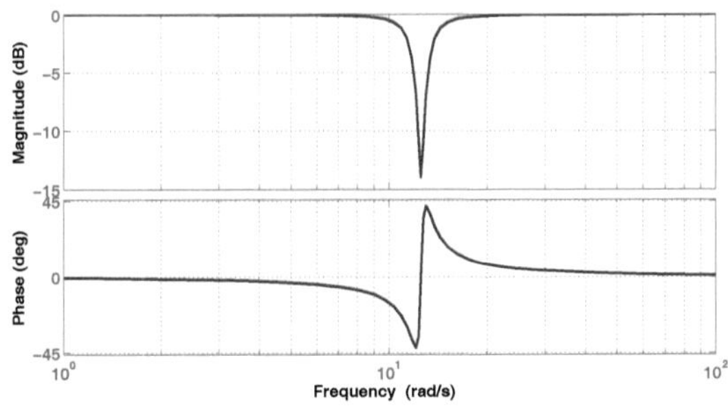

FIGURE 2.9: Bode diagram of an exemplary notch filter

In case of no active baseline controller being present, a static trim at takeoff is added to the actuator position and does not change during flight, i.e. the actuators are in some position near the center when the $\mathcal{L}_1$-controller is initialized.

Integrators are equipped with an anti-wind-up mechanism as shown in Figure 2.10, where $k < 0$.

This module is important to prevent the integrator value from exceeding the input channel saturation bounds, and with conservatively chosen integration limits to leave control authority to the "P" and "D" terms – and if present – the augmenting controller. Helicopters exist, wherein the integrator is active in trimmed flight only, cf. [10].

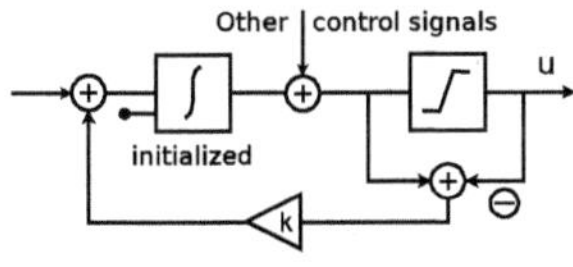

FIGURE 2.10: Anti-wind-up module

None of the gains or parameters is scheduled over the envelope; if performance is regarded poor, decoupling functions may be scheduled first.

Turn Compensation and Coordination

With the Euler angles Φ, Θ, Ψ, and the body-fixed turn rates p, q, r, the strap-down equation is:

$$\begin{bmatrix} \dot{\Phi} \\ \dot{\Theta} \\ \dot{\Psi} \end{bmatrix} = \begin{bmatrix} 1 & sin(\Phi)tan(\Theta) & cos(\Phi)tan(\Theta) \\ 0 & cos(\Phi) & -sin(\Phi) \\ 0 & \frac{sin(\Phi)}{cos(\Theta)} & \frac{cos(\Phi)}{cos(\Theta)} \end{bmatrix} \begin{bmatrix} p \\ q \\ r \end{bmatrix} \tag{2.3}$$

Its inverse form is given by:

$$(\vec{\omega}^{OB})_B = \begin{bmatrix} p \\ q \\ r \end{bmatrix} = \begin{bmatrix} 1 & 0 & -sin(\Theta) \\ 0 & cos(\Phi) & sin(\Phi)cos(\Theta) \\ 0 & -sin(\Phi) & cos(\Phi)cos(\Theta) \end{bmatrix} \begin{bmatrix} \dot{\Phi} \\ \dot{\Theta} \\ \dot{\Psi} \end{bmatrix} \tag{2.4}$$

During a coordinated turn in a horizontal plane, the lateral acceleration relative to the fuselage is zero. Hence:

$$Lcos(\Phi) = mg; \quad Lsin(\Phi) = mV\dot{\Psi} \tag{2.5}$$

where L is the lift in body-fixed coordinates, m is the helicopter mass, and g the local gravity.

From Equation (2.5) it follows that $\dot{\Psi} = cos(\Phi)\frac{g}{V}$ for a coordinated turn, and together with the assumption of a steady state turn, i.e. $\dot{\Phi} = 0$, $\dot{\Theta} = 0$, the compensated turn rates $\omega_{k,comp}$ depicted in Figure 2.5 and Figure 2.7 are:

$$p_{comp} = -sin(\Theta)cos(\Phi)\frac{g}{V} \tag{2.6a}$$

$$q_{comp} = sin^2(\Phi)\frac{cos(\Theta)}{cos(\Phi)}\frac{g}{V} \tag{2.6b}$$

$$r_{comp} = sin(\Phi)cos(\Theta)\frac{g}{V} \tag{2.6c}$$

This defines the compensated turn rates: $\omega_{k,turn} = \omega_k - \omega_{k,comp}$ with $\omega_k \in \{p, q, r\}$.

The roll rate during a coordinated turn is expected to be small and hence omitted in the implementation.

Turn coordination is switched on if velocity and bank angle exceed some threshold. It may be deactivated e.g. for aerobatics flight by a manual switch.

To eschew a surge in the turn compensation signals in case of flying with large bank angles and reaching the velocity threshold from inside, the turn rate signal is low-pass filtered after the switching structure so that steps in the turn command are attenuated.

Controller Tuning

Tuning of the baseline controller is done in frequency domain by loop shaping. For the basics of loop shaping, read e.g. [11], chapter 11. A more general overview about control authority, robustness, and frequency responses can be found in [12]. Details about this particular baseline controller are described in [3].

Initially, only the feedback loop is tuned. Afterwards, the feedforward control is shaped to obtain a closed-loop DC-gain of one without any overshoot for the achievable bandwidth.

For gathering information about the helicopter, system identification is applied as described in section 2.3. The open-loop frequency response is used in the form of a data set to be multiplied with the controller transfer function:

$$BSL(s) = \frac{1}{s} \left(\frac{K_i}{s} + K_p + K_d s \right) LL(s) \tag{2.7}$$

where K_p, K_i, and K_d are the "PID" gains, $LL(s)$ are lead-lag filters introduced in Equation (2.1).

Then the loop transfer function $BSL(s) \cdot plant(s)$ is obtained. The "*plant*" is defined as dynamics between command for the actuators and angular rates, i.e. the bare airframe model together with the actuators. This is the input output pair used for system identification to tune the feedback controller. The transfer function in (2.7) would not be proper and thus could not be utilized in any analysis without the additional integrator. Hence, Equation (2.7) contains a supplemental $1/s$, i.e. the measured rate is transformed to attitudes for feedback tuning.

Not included in this description are notch filters as well as trim and decoupling functions, which are added later (e.g. after controller tuning) in the topology of the controller.

Closed-loop tuning of the feedforward controller is based on the relation "actuator $\leftrightarrow$ measured attitude", analyzed in system identification.

In a linear SISO system, from the loop transfer function:

$$L(s) = BSL(s) \cdot plant(s) \tag{2.8}$$

the closed-loop transfer function $M(s)$ can be obtained as:

$$M(s) = \frac{L(s)}{1 + L(s)} \tag{2.9}$$

and the sensitivity function $S(s)$ as:

$$S(s) = \frac{1}{1 + L(s)} \tag{2.10}$$

The loop transfer function $L(s)$, the closed-loop "desired" system $M(s)$ and the sensitivity function $S(s)$ are used for loop shaping. Various plots can be used to support the manual loop shaping procedure, including the loop transfer function with the gain and phase margins, the sensitivity function showing the disturbance rejection dependent on the frequency, and the Nichols diagram to complete the picture, namely to hide the frequencies in the axes of the sensitivity function. The Nichols diagram was more important in times prior to computer aided tuning, but still helps the engineer in the design process. The analytically computed sensitivity function (2.10) can be verified by applying a frequency sweep to the actuators, acting as input disturbance. Note that in linear SISO systems, input-sensitivity and output-sensitivity functions are equivalent.

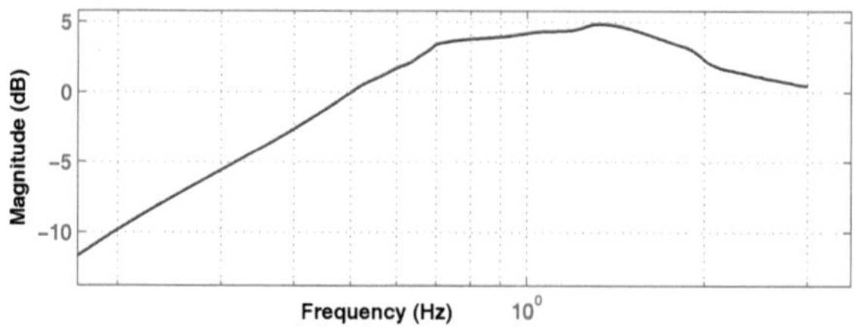

FIGURE 2.11: Closed-loop feedback tuning in the pitch axis – sensitivity function

Figures 2.11, 2.12 and 2.13 are obtained at 60 *kts* straight level flight, MTOM and aft limit of CG at low altitude ('Gm' = "Gain margin", 'Pm' = "Phase margin").

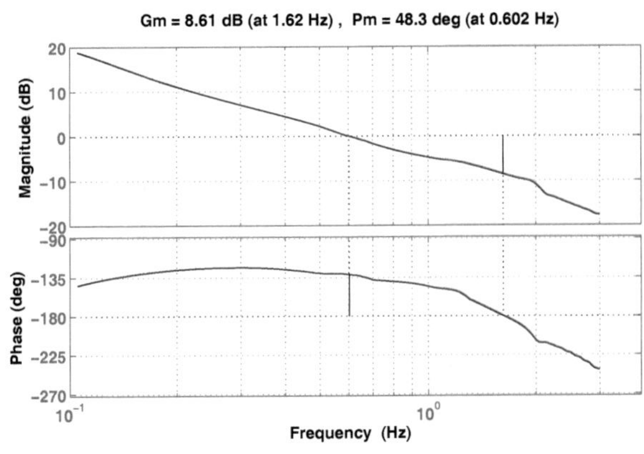

FIGURE 2.12: Closed-loop feedback tuning in the pitch axis – loop transfer function (bode plot)

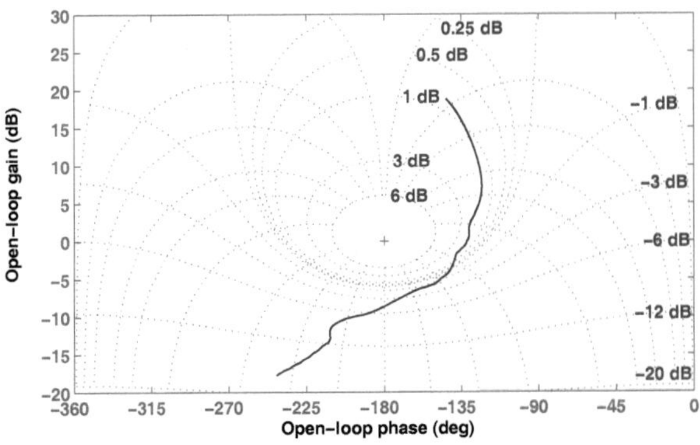

FIGURE 2.13: Closed-loop feedback tuning in the pitch axis – Nichols diagram

These curves are designed towards the following objectives: Curves shall be smooth. The peak of the sensitivity function should be kept small (e.g. below 5 *dB*), but in a trade-off with other frequencies – the integral of the sensitivity function is subject to conservation laws, cf. the Bode-Integral e.g. in [12]. A shift of the peak to higher frequencies means faster disturbance rejection. The amplitude of the loop transfer function in some cases is decreased in high frequencies in case of excessively high feedback of measurement noise. Specifications usually require minimum margins of 6 *dB* and 45 *deg*.

Due to the large number of parameters to be tuned, manual tuning of such a controller requires some experience. Most challenging is the trade-off between multiple objectives – a multi-criteria optimization with a user defined Pareto front. Hence, the author describes in [3] an automated approach: A combination of constrained nonlinear optimization algorithms, namely a sequence of quadratic programming together with evolutionary algorithms is applied to optimize the frequency responses to a predefined desired frequency response, which can be modeled as low order transfer functions. The sequence of quadratic programming requires some form of line search, e.g. the golden section method. Either one desired frequency response alone or a transfer function for every diagram can be defined as an optimum in nonlinear optimization. The latter case minimizes a cost function composed of several secondary cost functions, one for every diagram and without ever reaching zero in the total cost function. Nonlinear cost functions (e.g. the sum of the distances in every frequency response squared) ensure *smooth* curves.

With the feedback controller being accomplished, the feedforward controller is tuned. This includes the attitude and rate feedforward gains, and filters if required. The proportional feedforward gain can be used to tune low frequencies, i.e. it shifts the response to a higher or lower amplitude. The differential gain applies mostly to higher frequencies, for which filters (e.g. low-pass or lead-lag) can be utilized, too.

Both, feedforward as well as feedback tuning, are usually conducted iteratively – independent of a manual or automatic tuning strategy. After one tuning step, system identification is performed with the new feedback controller, which in turn is retuned with the new data. Nonlinearities and identification errors are thus reduced gradually. This does not contradict the recommendation of section 2.3 to perform system identification with a *deactivated* baseline controller, as an accurate open-loop model is not required anymore, rather the closed loop performance is of interest.

Finally, the controller is flight tested over the envelope; a controller tuned at 60 *kts* straight level flight does not guarantee satsifying performance in autorotation or fierce turns. It may have to be retuned or slightly adjusted.

Note that this methodology as well as the requirements stated in [4] imply a linear system, which in practice is a justified assumption – at least for this application.

Loop shaping is known and applied beyond helicopter design; new forms of controllers and tuning however exist.

Design of Decoupling Functions

There are two kinds of cross-couplings: Rate and command couplings. Decoupling functions, which ares mixing signals of different axes, are designed similarly for both kinds.

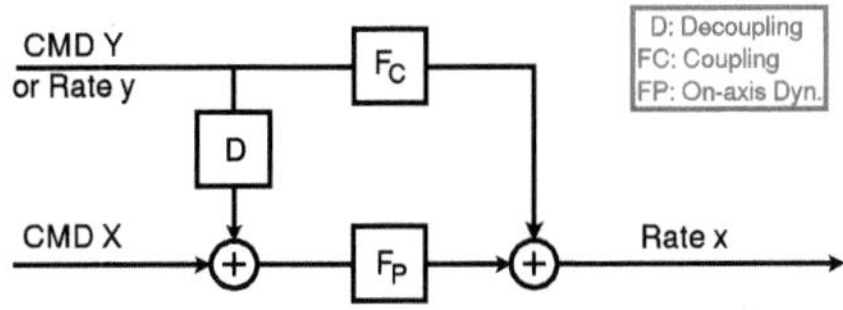

FIGURE 2.14: Coupling scheme – designing a decoupling function

In Figure 2.14, F_C and F_D are natural (system-inherent) dynamics, and D is designed such that:

$$F_C + F_P D = 0 \tag{2.11}$$

I.e., D cancels the effect of "CMD Y" or "Rate y" on "Rate x", since "CMD X" is desired to be the only cause of "Rate x", without any influence from other axes, neither from commands nor rates.

Designing D requires sufficient knowledge of the dynamics F_C and F_P.

In the following applications of system identification, the frequency sweep to excite dynamics is applied to the input, placed in the denominator of each fraction $\frac{out}{in}$.

F_P is identified by $\frac{x}{X}$.

With $Y F_C + X F_P = x$ and $y F_C + X F_P = x$, respectively,
F_C is identified by $\frac{x}{Y} - \frac{X}{Y} F_P$ for command decoupling or $\frac{x}{y} - \frac{X}{y} F_P$ for rate decoupling.

In case of rates, i.e. "y", the identification from a causal rate to an output rate is not very accurate due to small signals and a small signal to noise ratio. For this reason, $\frac{x}{y}$ is obtained by $\frac{x}{Y} \frac{1}{\frac{y}{Y}}$, and $\frac{X}{y}$ by $\frac{X}{Y} \frac{1}{\frac{y}{Y}}$ so that the input is always a frequency sweep. See also [3].

2.5 Introduction to $\mathcal{L}_1$-Control

$\mathcal{L}_1$-control refers to a class of controllers with frequency scale decoupling of adaptation and control loops. The name is derived from the integral of impulse responses of proper and BIBO stable transfer functions, cf. [13].

2.5.1 The Idea

The purpose of faster time scales in information processing than in physical control is to provide the "right" control signal as fast as possible. The same idea is used e.g. in feed-forward dynamic inversion to cancel unwanted dynamics before they appear in a tracking error signal.

This idea is realized – regardless of the specific form – with the following elements:

The predictor is a software unit that runs in parallel to the real plant dynamics, representing some form of desired dynamics.
The control law with a filtering structure primarily ensures cancellation of the uncertainties with low-frequency content. Second, it becomes part of a resulting high-pass filter for adding high-pass uncertainties to the desired dynamics (i.e. to the predictor).
The adaptive law provides a term based on the uncertainty estimate.

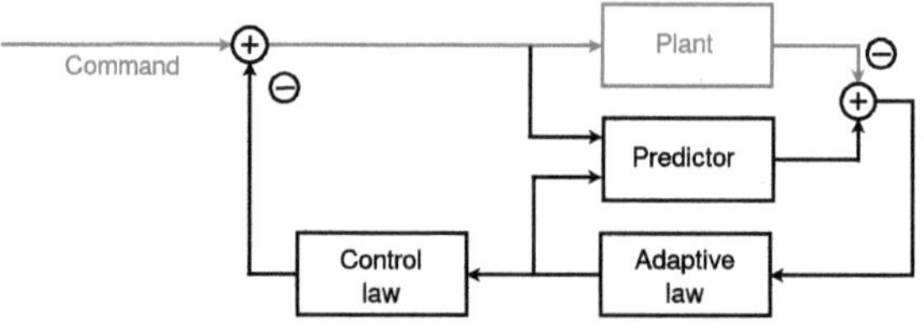

FIGURE 2.15: Simplified concept drawing of the $\mathcal{L}_1$-controller structure

The basic elements of an $\mathcal{L}_1$-controller are shown in Figure 2.15. It remains to be seen whether in future architectures different concepts with fast adaptation can be utilized.

As in any other control system, desired dynamics should be chosen feasible or nearly feasible for sufficient robustness.
When derived from a dynamic model, commands can only be perfectly tracked if this dynamic model has at least the same relative degree as the plant [14].

The elements are now discussed in detail:

The Control Law

In any mechanical control problem, the input channel bandwidth, i.e. control authority is limited. In such mechanical systems, this bandwidth is determined by the actuators. In addition, the response is determined by the inertia of the plant.

To account for this limitation and to prevent high frequencies from the fast adaptation to propagate to the plant, the control law has a filtering structure with low-pass characteristics. Its bandwidth and order is defined by design, ideally in accordance with the actuator bandwidth. This is one of the elements in the trade-off between performance and robustness and describes "aggressiveness" of control, with its effects on actuator wear.

At the same time, it indirectly acts as part of a complimentary filter structure (while only one filter needs to be implemented) to add the high frequency content to the desired dynamics.

Let the predictor be $\dot{\hat{x}}(t) = a\hat{x}(t) + b(u(t) + \hat{\sigma}(t))$ and the input $u(s) = -C(s)\hat{\sigma}(s)$, where $C(s)$ is a low-pass filter and a and b state the desired linear dynamics, $\hat{\sigma}(t)$ is a term based on the cumulated uncertainty estimate, then the closed-loop predictor is in time domain:

$$\dot{\hat{x}}(t) = a\hat{x}(t) + \eta(t); \quad \eta(t) = \mathscr{L}^{-1}\left\{b(1 - C(s))\hat{\sigma}(s)\right\} \tag{2.12}$$

In this equation, $\mathscr{L}^{-1}\{\cdot\}$ describes the transformation from Laplace to time domain. The so originated high-pass filter $(1 - C(s))$ serves the purpose of actively deteriorating the desired system with respect to high frequencies in the control signal, that are assumed to have no beneficial effect on the stability of the system, but are hurting robustness. Adding high frequency uncertainties shifts the role of the predictor from stating solely desired dynamics to a predictor as such. The control signal is applied to the plant and the predictor at the same time, rendering the prediction error close to zero.

The high-pass filtered estimated uncertainties affect $\dot{\hat{x}}(t)$, which gets integrated to $\hat{x}(t)$. Equation (2.12) in frequency domain is:

$$\hat{x}(s) = \frac{b}{s - a}(1 - C(s))\hat{\sigma}(s) \tag{2.13}$$

The low-pass nature of the control signal combined with the uncertainties affected by a high-pass approximate a no-pass filter for uncertainties to $\hat{x}(s)$. Hence, the uncertainties have little effect on $\hat{x}(t)$. By either lowering the bandwidth of the desired system or by increasing the bandwidth of the low-pass filter, the effect of the cumulated uncertainties on the state prediction is reduced. This is consistent with the later introduced reference system (a virtual closed-loop plant for which the uncertainties are perfectly known), whose stability condition relies on a sufficiently small $\left\|\frac{b}{s-a}(1 - C(s))\right\|_{\mathcal{L}_1} L$ – valid for this example and where L is a Lipschitz constant for the uncertainties/disturbances. Increasing the bandwidth of a first order low-pass filter for instance lowers this $\mathcal{L}_1$-norm. The costs are higher frequencies in the input channel – an "optimum" is emerging. (An optimum is what is defined to be one.)

In addition, the filter is an instrument of noise attenuation. Due to the high gain nature of the adaptation loop, the measurement noise enters unfiltered into the loop "predictor - adaptive law". In this loop, it is isolated from the control signal and harmless to it. Depending on the kind of adaptive law, the relative degree of the filter can be crucial. With the later explained piece-wise constant adaptive law, this is not an issue, as later explained.

A filter in the control signal introduces phase lag, similar to a PI-controller. Hence, control lags behind the identified uncertainties already in the controller architecture. This may be considered a drawback compared to some other predictor based architectures without filtering. However, in these applications the trade-off between performance and robustness is often primarily attributable to the adaptive gain, which in $\mathcal{L}_1$-control could be as high as hardware limits allow (without any negative impact on robustness), as will be appreciated from a later analysis and e.g. [13]. Shifting this trade-off to the filter enhances the design and enables locally guaranteed transient performance.

According to this, it is not the purpose of $\mathcal{L}_1$-control to compensate with high frequencies for disturbances and uncertainties, but to capture those fast and send the "right" low-frequency control signal ab initio.

If notch filters for the control signal are desired to avoid excitation of structural modes, these filters can be integrated in the control law, i.e. in the low-pass filter structure. Recall that in the baseline controller notch filters are applied at the baseline controller output.

ADS-33 ([4]) suggests that requirements formulated in frequency domain are often more suitable for helicopters than those in time domain, exceptions exist. This is consistent with the fact that to this day most helicopter controllers are designed in frequency domain. Designing an $\mathcal{L}_1$-controller is basically different to this as the $\mathcal{L}_1$-controller is (internal) model based. With the introduction of the filter however, this gap narrows and the frequency domain is revitalized.

The Predictor
By splitting the plant dynamics into

a) known, desired dynamics and

b) in general unknown, unwanted dynamics and disturbances,

the predictor contains desired dynamics explicitly, hence the category "model following control". The predictor defines the desired dynamics as e.g. state space model or transfer function, to which a fast estimation of uncertainties is added. It is the control goal to cancel these uncertainties unless outside the input channel bandwidth, such that desired dynamics remain apart from uncertainties that lie outside the filtering bandwidth. Plant and predictor thus have approximately the same state vector and output, respectively. The difference is called prediction error. This error is a measurement for controller quality.

The short term dynamics, i.e. (angular) rates of the vehicle in reaction to the actuator inputs, are the substantive relation for predictors in this thesis. On the contrary, any of the characteristics regarding power requirements or endurance is meaningless for an inner loop controller. Either nominal dynamics obtained by system identification or dynamics akin to it – usually a faster version and with an aperiodic step response of them – are defined as desired behavior in the predictor. The slight non-minimum phase nature of the helicopter can be bypassed with additional modeled time delay in the input channel to the predictor. In conclusion, a first order system from actuator input to angular rates with the proper amount of modeled time delay represents an appropriate strategy. It is verified in simulations and finally flight tests.

Angular rates are to be preferred to angular accelerations due to the poor signal quality of the latter on helicopters. Attitude angles are disregarded as the dynamic uncertainties are fully captured with the input to angular rates since no uncertainties between angular rates and attitude angles exist; in other words, the strap-down equation (2.3) is uncertainty free; over-determining the uncertainty estimation is avoided then. The relation is a fully kinematic one in contrast to e.g. the one between AoA and pitch rate. Additionally, attitude signals are provided usually with a higher latency, which would harm simultaneous adaptation.

Regardless of being a state or output predictor, states or an output are predicted by virtue of information in the input channel – an input predictor however could be employed, too. The explicit use of the current input signal allows for hedging of input saturations, time delay, actuator dynamics and everything else affecting the system input unless added later to the signal. This is the analog of anti-wind-up modules of PI-controllers. If the input of the plant saturates, the same saturation can take place in the predictor input and hence limit the input to the desired dynamics – the prediction error remains small. The hedging also means that systems with slow actuators are treated equally, only the overall system performance suffers due to missing control authority.

The implemented predictor consumes CPU-power. On the other hand, it is to state a feasible control goal with desirable performance properties, thus possibly requiring complex relations. This trade-off between information content and CPU load suggests state space models or transfer functions in the first instance, both as simple as possible.

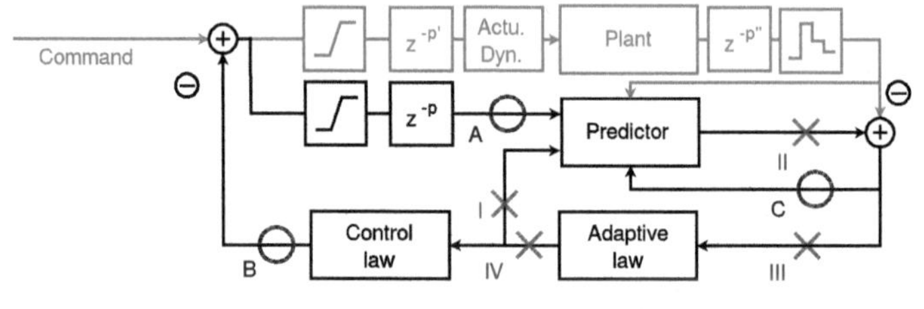

FIGURE 2.16: Detailed concept drawing of the $\mathcal{L}_1$-controller structure

Figure 2.16 shows a detailed structure of an $\mathcal{L}_1$-control architecture, however without any hedging structures, cf. [15]. The exogeneous input (e.g. from the baseline controller) does not contain a trim signal that would be hedged from the predictor in this figure, cf. Figures 3.1 and 3.2. With hedging structures, the modeled saturation shown in Figure 2.16 is abundant due to the natural compliance in the hedging structure.

A zero-order-hold block at the output of a continuous plant represents a discrete sensor rate. The spots *I...IV* usually remain unmodified to ensure integrity of the adaptation loop. This applies to modeled sensor time delays, which should not be modeled at the predictor output – at least for the concepts shown in this thesis. The sensor time delay (which includes delays in buses and sensor unit processors if in place) may be included to some extent in the modeled time delay of the input channel. It can be expected to be much smaller than the delay of the input channel.

Note that Figure 2.16 is intended for the piece-wise constant adaptive law, a projection operator to prevent parameter drift would interfere at this point. Projection operators however are not used in this thesis as parameter drift is not a potential issue (due to *one* cumulated estimation instead of at least two parameters, which potentially drift apart and mutually more or less cancel).

It is up to the engineer how to use the output of the adaptive law for the control law – e.g. to limit it. At spot *A*, the input channel can be resembled. At spot *B*, the adaptive input can be limited, e.g. in augmentation. Spot *C* allows for modified feedback of the prediction error. Similar statements are presented in [15].

Furthermore, the predictor parameters together with the adaptive law and if necessary the control law can be scheduled.

The Adaptive Law

Depending on the adaptive law, different terms for different kinds of uncertainties are introduced, e.g. state or input dependent uncertainties or external disturbances. In some adaptive laws, all uncertainties are lumped into one variable.

The following adaptive laws can be used in $\mathcal{L}_1$-control, shown with potential arguments in favor "+" and against "−" the respective approach, but without claim to completeness.

Gradient minimization type. Based on Lyapunov theory, this adaptive law acts as nonlinear integrator since the derivatives of parameters are updated; often used with the projection operator to provide the required robustness modification. This kind of adaptive law is not used in this thesis.
- Computationally expensive.
- Designed for continuous systems (including sensor and CPU).
- Requires modification against parameter drift.
+ Potential of including more than one parameter can be desirable.

Piece-wise constant adaptive law. This adaptation law takes the CPU time step into account. A specific proportional gain is applied to the prediction error. By inverting the error dynamics, it drives the prediction error close to zero after every time step. Modifications exist to utilize earlier time steps rather than only the last one, to "learn" the uncertainty.
+ Includes explicitly the computational time step length.
+ Modest computational effort.
+ Avoids the need of robustness modifications, e.g. projection operator.
+ The gain applied to the prediction error can (usually) be lower than what would be needed for the gradient based laws.

Proportional adaptive law. This law is introduced in [16]. It consists solely of a gain as free design parameter applied to the prediction error.
- Designed for continuous systems (including sensor and CPU).

All laws lead to performance bounds of the same structure, namely proportional to some function that includes the inverse adaptive gain (for continuous systems) or the CPU time step length (for the piece-wise constant adaptive law), cf. e.g. [13].

For all kinds of adaptive laws, the gain margin regarding the adaptive gain is infinity (cf. [13]), wherein in the piece-wise constant law the gain is additionally influenced by the time step length. As inversion of the error dynamics without an explicit parameter estimation, it also includes the sampling time for example and is the signal to drive the prediction error close to zero.

Besides these obvious advantages of the piece-wise adaptive law, it enables a structure in output feedback that allows to overcome the limitations introduced with e.g. the Kalman-Yakubovich Lemma ("$Pb = c$"), i.e. this theory does not require the desired dynamics to be SPR (strictly positive real), cf. e.g. [13].

2.5.2   Explanation to the Piece-Wise Constant Adaptive Law

Assume the following scalar plant dynamics, where a_P is the plant system matrix, b the known input matrix, ω the unknown input gain, $f(x(t), t)$ the unknown time varying, state dependent nonlinearity, and $d(t)$ the unknown time varying external disturbance:

$$\dot{x}(t) = a_P x(t) + b\omega u(t) + f(x(t), t) + d(t) \tag{2.14}$$

Then, by lumping all uncertainties into one variable $\sigma(t)$:

$$b\sigma(t) = (a_P - a)x(t) + b(\omega - 1)u(t) + f(x(t), t) + d(t) \tag{2.15}$$

one has:

$$\text{Plant: } \dot{x}(t) = ax(t) + b(u(t) + \sigma(t)) \tag{2.16}$$

$$\text{State predictor: } \dot{\hat{x}}(t) = ax(t) + b(u(t) + \hat{\sigma}(t)) + a_{SP}\tilde{x}(t) \tag{2.17}$$

$$\text{Error dynamics: } \dot{\tilde{x}}(t) = a_{SP}\tilde{x}(t) + b\hat{\sigma}(t) - b\sigma(t) \tag{2.18}$$

where $a < 0$ and b are desired values of linear first order dynamics, which state the control goal. The error dynamics are shaped by $a_{SP} < 0$. The idea of assigning poles of the error dynamics different to the desired dynamics is first presented in [17]. See also Appendix D for clarification.

The goal is to minimize the prediction error $\tilde{x}(t)$. The adaptive law is such that $\hat{\sigma}(iT)$ drives $\tilde{x}((i + 1)T)$ as close to zero as possible, with T being the computational time step length and $i = 0, 1, 2, \dots$.

By integrating the differential equation of the error dynamics (2.18) from iT to $(i + 1)T$, one has:

$$\tilde{x}((i + 1)T) = e^{a_{SP}T}\tilde{x}(iT) + \int_0^T e^{a_{SP}(T-\tau)}b\hat{\sigma}(iT)d\tau - \int_0^T e^{a_{SP}(T-\tau)}b\sigma(iT + \tau)d\tau \quad (2.19)$$

as $\sigma(iT)$ is unknown but not $\tilde{x}(iT)$, the best to be done so far is to find a $\hat{\sigma}(iT)$ to cancel the first two terms. During the integration step, $\hat{\sigma}(iT)$ is a constant. Hence,

$$\hat{\sigma}(iT) = -\frac{1}{b}\left(\int_0^T e^{a_{SP}(T-\tau)}d\tau\right)^{-1}e^{a_{SP}T}\tilde{x}(iT) = \frac{a_{SP}e^{a_{SP}T}}{b\left(1 - e^{a_{SP}T}\right)}\tilde{x}(iT) \quad (2.20)$$

The remaining term is:

$$\tilde{x}((i + 1)T) = -\int_0^T e^{a_{SP}(T-\tau)}b\sigma(iT + \tau)d\tau \quad (2.21)$$

The intuitive explanation for this is: The controller applies an adaptive parameter such that the error dynamics, which accumulated through uncertainties in the *previous* time step, is driven to zero in the *next* time step. During that interval, while a $\hat{\sigma}(iT)$ is "working" to render $\tilde{x}((i + 1)T)$ zero, the uncertainty accumulates again to a new $\tilde{x}((i + 1)T)$. This can be seen as the remaining term in Equation (2.21). The shorter T, the smaller is $|\tilde{x}(t)|$. Hence, the choice of T is significantly influencing closed-loop system performance.

Note that this is not the control law. The purpose of the control law is *not* to minimize $\tilde{x}(t)$, but to cancel $\sigma(t)$.

> The variable $\hat{\sigma}(iT)$ is not exactly an estimate of $\sigma(t)$ as it also includes the time step T.

The variable $\hat{\sigma}(iT)$ is the value to cancel the first terms of the error dynamics after every iT. Due to this fact, any additional robustness modifications are not required.

The remaining term causes the prediction error to be bounded and even small, but not to converge asymptotically to zero. With this in mind, different predictors or adaptive laws can be envisaged. It is to be decided case by case, whether a modification is needed or the sampling time is sufficient to obtain a well performing $\tilde{x}(t)$.

In any case, the underlying assumption is that T is far below the time scales which $\sigma(t)$ is changing with – an assumption that is well justified in most systems. This assumption is also important for the fact that sensor and CPU delays add a time offset to this mechanism. The assumption is expanded that this offset is below any significant change of $\sigma(t)$ during the offset. This is usually true as the offset is expected to be close to T.

These assumptions (change rate of uncertainties slower than update mechanism) imply that some learning mechanism would be useful that remembers the remaining term or the uncertainty, in order to add this information either to the predictor or directly the adaptive law.

A recursive adaptive law is introduced next.

Recursive Piece-wise Constant Adaptive Law

This law is first introduced in [18]. It modifies the adaptive law (2.20) towards:

$$\hat{\sigma}(iT) = \frac{a_{SP}e^{a_{SP}T}}{b\left(1 - e^{a_{SP}T}\right)}\tilde{x}(iT) + \frac{-a_{SP}}{b\left(1 - e^{a_{SP}T}\right)}h(iT) \tag{2.22}$$

where:

$$h(iT) = -\tilde{x}(iT) + h((i-1)T); \quad h(0) = 0; \; i = 1, 2, ... \tag{2.23}$$

The recursive term introduces integral character. Asymptotic convergence of $\tilde{x}(t)$ is possible with this adaptive law, *iff* $\sigma(t)$ is constant, see Appendix E.5. By taking information from the past, it resembles "learning" despite working with fast adaptation. This adaptive law is thus capable of reducing $\tilde{x}(t)$ significantly.

It can be shown in simple simulations that this adaptive law has the same effect has decreasing T in the non-modified version. This advantage is not always possible to be exploited since this modification may be observed to reduce the time delay margin. This observation is confirmed in [19].

2.5.3 Introduction to Output Feedback for Non-SPR Desired Dynamics

Output feedback takes internal dynamics actively into account. A form of uncertainty estimation is conducted for states whose measurement is not available. State and output feedback may use the same input-output pair though, i.e. actuator signal and angular rates. Purely the explicit consideration of internal states renders the measured state (angular rate) an output.

Among a myriad of theories, the following is chosen to be the primary architecture in this thesis. It is shown e.g. in [13]. Although developed for *linear* uncertain plants with output dependent nonlinearities, it is considered well suitable for the stated problem: First, output dependent nonlinearities together with disturbances are expected to be a big portion of all uncertainties. Second, the fast adaptation leads to a fast approximation of a linear plant with disturbances. Any continuous function can be approximated by a piece-wise constant function with sufficiently small time step lengths – especially for a helicopter, where measurement noise is assumed to be larger than the information content of state dependent nonlinearities in the measurement.

$$\text{Plant:} \quad y(s) = M(s)(u(s) + \sigma(s))$$

$$\text{Predictor:} \quad \dot{\hat{x}}(t) = A\hat{x}(t) + bu(t) + \hat{\sigma}(t), \quad \hat{x}(0) = \hat{x}_0; \quad \hat{y}(t) = c^T \hat{x}(t)$$

$$\text{Control law:} \quad u(s) = C_r(s)r(s) - C(s)\frac{M_{um}(s)}{M(s)}\hat{\sigma}(s)$$

$$\text{Adaptive law:} \quad \hat{\sigma}(iT) = -\Phi(T)^{-1}\Upsilon(T)1_1\tilde{y}(iT) + \Phi^{-1}(T)1_1 h(iT), \quad i = 0, 1, 2, .$$

where A, b, c^T is a minimal state space realization of the desired dynamics $M(s)$; $C(s)$ and $C_r(s)$ are strictly proper and stable low-pass filters that are initialized with zero. Furthermore, $1_1 = [1, 0, ..., 0] \in \mathbb{R}^n$, and

$$h(iT) = -\tilde{y}(iT) + h((i-1)T), \quad h(0) = 0, \quad i = 1, 2, ...$$

$$\Lambda = \begin{bmatrix} c^T \\ D\sqrt{P} \end{bmatrix}, \text{with } D \text{ such that: } D\left(c^T\sqrt{P}^{-1}\right)^T = 0$$

$$\text{and } P \text{ such that } A^T P + PA = -Q$$

$$\text{and } Q \text{ as design element such that } Q = Q^T > 0$$

$$\Phi(T) = \left(-\Lambda A\Lambda^{-1}\right)^{-1}\left(\mathbb{I} - e^{\Lambda A\Lambda^{-1}T}\right)\Lambda \qquad M(s) = c^T(s\mathbb{I} - A)^{-1}b$$

$$\Upsilon(T) = e^{\Lambda A\Lambda^{-1}T} \qquad M_{um}(s) = c^T(s\mathbb{I} - A)^{-1}$$

A detailed explanation is given in Chapter 4.3, an analysis is conducted in Appendix F.

Variations can be found: E.g. an *input* predictor in [20], a system for linear time-varying desired dynamics in [21], or for state dependent nonlinearities in [22].

2.5.4 Performance Effects of the Prediction Error: Issues and Solutions

The terminology used in this chapter mostly refers to state feedback, but the effects are prevalent also in output feedback.

In the formal proofs of Appendices E and F it is shown that $\tilde{x}(t)$ and $\tilde{y}(t)$, respectively in general lack asymptotic convergence to zero but converge close to it.

As later shown by Equations (H.13) and (E.68) on pages 183 and 121, namely:

$$\tilde{x}((i_0 + 1)T) = -\int_0^T e^{a_{SP}(T-\tau)}b\sigma(i_0 T + \tau)d\tau$$

$$\|(x_{ref} - x)_i\|_{\mathcal{L}_\infty} \leq \frac{\|(s-a)^{-1}bC(s)(s-a_{SP})/b\|_{\mathcal{L}_1}}{1 - \|(s-a)^{-1}b(1-C(s))\|_{\mathcal{L}_1} L}\|\tilde{x}_i\|_{\mathcal{L}_\infty}$$

a $\sigma(t)$ causes a nonzero prediction error that propagates to the tracking error. These equations show potential tuning parameters to weaken the effects of $\sigma(t)$ on the tracking error.

For a position or attitude controller this means some deviation from the commanded position or attitude, which in most cases is negligible. For controllers however, wherein the commands describe angular rates, an offset from the commands causes drift. This is especially true if no outer loop controlling the position or attitude is present. This drift was initially problematic in this thesis when designing the vertical speed controller as standalone $\mathcal{L}_1$-controller.

As a, b are fixed by the desired dynamics and $C(s)$ is strongly oriented at the bandwidth the actuators offer, the equations suggest to work with $|\sigma(t)|$, T and a_{SP}:

Discussion of Potential Solutions

1. For state feedback with piece-wise constant adaptive law only: Choose a_{SP} slow, e.g. $(0.01...0.1)a$. The functionality remains by inversion of the error dynamics. For more details see Appendix H.

2. Apply a memorizing or integral term to "learn" $\sigma(t)$. This can reduce $\tilde{x}(t)$ significantly. In this thesis, a recursive adaptive law introducing integral behavior to the adaptive law is shown, cf. Equation (2.22) and Equation (2.23).

3. Reduce the time step length T for a shorter integration time of the new uncertainties. A hardware upgrade however is not always viable, especially in aerospace applications.

4. Apply a multirate controller. I.e., more than one adaptation loops for one sensor step length are applied. This is similar to reducing T but bypasses the requirement for new sensor information.

5. Cancel known parts of undesired dynamics and disturbances by a baseline controller. Especially large (constant) disturbances canceled by a trim signal can reduce $\sigma(t)$ significantly. This is called augmentation.

Performance and robustness effects of the respective solutions are discussed next.

2.5.5 Performance and Robustness

Both, state and output feedback as shown in Appendix E and Appendix F allow internal dynamics; with the difference that state feedback is robust against it, whereas the shown output feedback controller actively accounts for internal dynamics in its estimations.

In **state feedback**, the trade-off between performance and robustness is reflected in the following design parameters:

- The bandwidth of the filtering structure (higher bandwidth is better performing, less robust).

- The order of the filtering structure (in general, lower order is better performing, less robust; zeros complicate the effects!).

- The choice of desired dynamics (in augmentation: any deviation from the real dynamics may be detrimental to robustness; for the standalone $\mathcal{L}_1$-controller: the slower the desired dynamics, the more robust).

- The amount of modeled time delay in the input channel (less delay is better performing, less robust – may be valid only in some region of time delay).

- The choice of a_{SP} (a slower a_{SP} in a scalar system is better performing, less robust, proof shown in Appendix H).

- The choice of the adaptive law (additional integral behavior may be better performing, less robust).

- The choice of the computational time step length T (shorter T is better performing *and* to some extent more robust – see proofs in Appendices E and F)

In **output feedback**, the trade-off is the same apart from a_{SP}, which is not existent in this form in output feedback.

> One of the points above hurting robustness is enough to end up with inadequate perfor-
> mance due to robustness issues. These issues become visible mostly in system oscillations,
> similar to limit cycles.

Bandwidth, a_{SP}, and modeled time delay are design parameters for the trade-off robustness and performance without impact on hardware requirements.

The recursive adaptive law consumes slightly higher computational effort per time step due to the additional operation, but in general allows for longer sampling times, thus reducing hardware requirements. This law however can have a negative impact on the time delay margin (cf. page 34) and these performance losses losses due to robustness issues may exceed performance gains from this modified adaptive law. Figure 2.17 illustrates this trade-off, where shorter T (not shown in the figure) have positive effects on performance and robustness while increasing hardware requirements (cf. Appendix E).

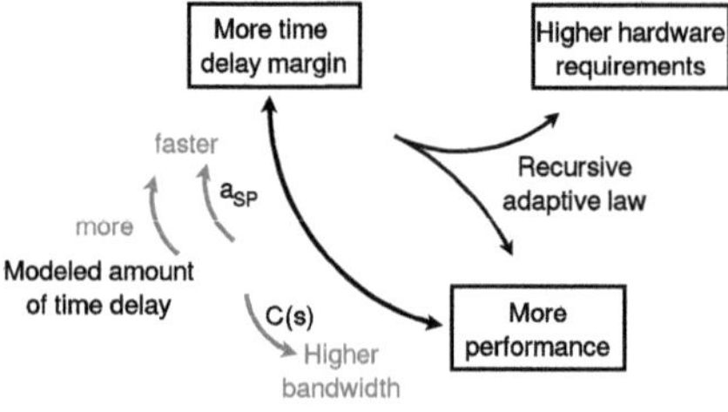

FIGURE 2.17: Elements of the trade-off: performance - robustness

2.6 Augmenting and Standalone $\mathcal{L}_1$-Controllers

Three options are envisaged:

1. Standalone $\mathcal{L}_1$-controller

2. Augmenting $\mathcal{L}_1$-controller (augmenting a baseline controller)

3. Switching to a more robust, (resilient) controller

In the following, these approaches are discussed in detail, with potential arguments in favor "+" and against "−" the respective approach, without claim to completeness.

Approach 1: Standalone $\mathcal{L}_1$-Control

The $\mathcal{L}_1$-controller is the only active controller. The defined desired dynamics resemble approximately the closed-loop dynamics "plant – baseline controller".

+ Closed-loop stability regarding sufficient compensation (later called the $\mathcal{L}_1$-norm condition without unmatched uncertainties) can be provided by either increasing the bandwidth of the filtering structure or by decreasing the bandwidth of the desired system. As the former is limited by maximum allowable actuator activity, the latter is to be chosen. This however is only viable with a standalone controller as in augmentation the desired dynamics have to be close to open-loop dynamics. (Valid for scalar systems without unmatched uncertainties)

+ It may be possible to achieve a higher reactive feedback. Despite the filtering structure, the standalone $\mathcal{L}_1$-controller is not bound to slow integrator dynamics that is further deteriorated by anti-wind-up strategies. In $\mathcal{L}_1$-control this functionality is already included in the input channel module.

− Less knowledge about the physical system than what is available is implemented. This is an disadvantage and advantage at the same time. On the one hand, the performance of a baseline controller in its design point may be hard to be surpassed with less system knowledge. On the other hand, either the baseline controller is away from the design point or so robustly designed that its full performance potential cannot be taken advantage of.

− For the sake of unique adaptation, the $\mathcal{L}_1$-controller is limited to either rates or attitudes.

Approach 2: Augmentation

> As shown in section 2.5.4, adaptive augmentation of the baseline controller is one of several options in the adaptive controller to reduce the propagation of cumulated uncertainties and disturbances to the tracking error. Additionally, other effects are introduced.

The system consists of a baseline controller for the "known" part and an adaptive component (here the $\mathcal{L}_1$-controller) for the "unknown". An example of applied augmentation is presented in [1] by the author, key ideas are repeated hereinafter.

The control objective of the $\mathcal{L}_1$-controller is to retain the nominal *open-loop* so that the baseline controller is working in its design point. In the same manner as proportional state feedback shifts the eigenvalues of a system, the $\mathcal{L}_1$-controller is used to retain nominal open-loop dynamics. Yet there is a difference – whereas the closed-loop eigenvalues of simple feedback control change with alternating open-loop eigenvalues, the $\mathcal{L}_1$-controller is stricter in its goal to always maintain the hard-coded desired dynamics (despite the inclusion of the limited control bandwidth), i.e. the eigenvalues remain constant if the control goal is fulfilled. The baseline controller is not limited to be a PI-controller but can e.g. be combined with a nonlinear dynamic inversion controller.

Another possible perception of the term "augmentation" would be: Preservation of the nominal *closed-loop* (closed by the baseline controller) dynamics. This however requires the same switching and saturation properties as introduced by the baseline controller. Hence, it is not pursued in this thesis.

The input to the predictor is the total input to the plant (trim and decoupling may be excluded from this). That is, saturation of the input signal is being taken fully into account.

The basic concept can be demonstrated in a simple example, where A_P (stable) is the plant system matrix and A_M (stable) defines desired dynamics, $u_{BSL}(t)$ is the baseline control signal, $u_{\mathcal{L}_1}(t)$ is the $\mathcal{L}_1$-control signal, $d(t)$ an external disturbance, and $r(t)$ some command.

$$\dot{x}(t) = A_P x(t) + B(u_{BSL}(t) + u_{\mathcal{L}_1}(t)) + d(t) \tag{2.24}$$

Let $u_{BSL}(t)$ be defined as $u_{BSL}(t) = -Kx(t) + r(t)$ and A_M as $A_M = A_P - BK$, then

$$\dot{x}(t) = A_m x(t) + B(u_{\mathcal{L}_1}(t) + r(t)) + d(t) \tag{2.25}$$

Then, one state predictor formulation is:

$$\dot{\hat{x}}(t) = A_P \hat{x}(t) + B(u_{BSL}(t) + u_{\mathcal{L}_1}(t) + \hat{\sigma}(t)) \tag{2.26}$$

Or equivalently:

$$\dot{\hat{x}}(t) = A_M \hat{x}(t) + B(u_{\mathcal{L}_1}(t) + \hat{\sigma}(t) + r(t)) \tag{2.27}$$

The alternatives are equivalent. However, as the total control signal is preferable for saturation modeling, the predictor is implemented in the form (2.26). Saturation relates to the sum of the inputs $u_{BSL}(t) + u_{\mathcal{L}_1}(t)$ and cannot always be split straightforwardly in its summands due to the nonlinear nature of the saturation operator.

Note that these equations show that the baseline control signal contains the command and the shaping of the desired dynamics. Thus, it is architecturally equal (apart from hedged signals and the saturation structure) to sending the command $r(t)$ to a standalone $\mathcal{L}_1$-controller.

In this architecture, attitude control (i.e. attitude feedback and feedforward with integrators for the attitude error) can be seen as outer loop of cascaded attitude and rate loops, and would be included in $r(t)$ in this example.

In state feedback, baseline controller states can be added as shown in Appendix M. These states stem from integrals (in this case the integral of attitude errors) and dynamic filters.

Augmentation in general may have the following potential advantages "+" and shortcomings "−":

+   The baseline controller helps to reduce $\sigma(t)$, thus has the tendency of improving the tracking error. This can be understood from a plurality of equations in this thesis.

+   The baseline-controller can be used as fallback mode in early flight tests, as the augmenting $\mathcal{L}_1$-controller can be switched on and off as desired (with some constraints in output feedback).

−   Augmentation without modifying the baseline control architecture in the strict sense destroys functionality of the anti-wind-up architectures unless the adaptive input signal $u_{\mathcal{L}_1}(t)$ is sent to the anti-wind-up function, too.

−   Hints emerge sometimes that the augmentation has smaller time delay margins, see e.g. Figure M.3. A general prove cannot be provided at this point.

−   CPU load is the highest of all options.

−   Tuning is cumbered by the coexistence of two controllers, see also [23].

−   Compatibility of the two controllers during the transient may be hard to guarantee. Different reaction times or phase lags may result in contradicting signals, and the predictor computes predicted states with this input.

−   The predictor model combined with a baseline controller that is designed by loop shaping and that is of higher order introduces undesired behavior unless perfectly tailored to each other.

The last point is critical to the augmentation with the baseline controller shown. In state feedback, a first order predictor introduces errors as shown in Figure N.3, page 214. In output feedback as shown in this thesis, constraints on the desired dynamics (e.g. relative degree 1), the tuning effort and complexity hamper finding a solution.

For this reason, augmentation can be helpful or harmful, there are arguments in favor of each.

Approach 3: Switching to a More Robust Controller

In case of severe system degradation, switching to a more robust $\mathcal{L}_1$-controller for a resilient emergency recovery may be an option. This option presumes some automatic failure detection or possibility for manual switching. Preconditionally for this is a sufficient margin of the helicopter, e.g. altitude to recover safely. This idea is presented e.g. in [24] for nuclear reactors and in [25] for flight envelope protection.

The main purpose of this controller is stabilizing the plant to stay aloft as opposed to recovering performance. Similar dynamics are targeted, however at lower bandwidths and with additional envelope protections.

+ May turn out to be the most survivable design.

− Additional software to implement.

− Performance loss.

All three options require sufficient control authority to function, see also [26] for performance of $\mathcal{L}_1$-controllers in the presence of input saturation.

With the augmentation possibly requiring the removal of baseline controller integrators and the modification of anti-wind-up structures, a potential requirement of not modifying the baseline controller for augmentation purposes may not be met.

2.7   Internal Model Based Control

Achievable Desired Dynamics

The reactions on a reduced plant's input gain of a PID feedback controller and a model following controller ($\mathcal{L}_1$-control) are different: The PID-controller's effectiveness is reduced equally (as much as this gain is reduced), apart from an error in the integrator channel of the controller, that ensures steady state accuracy.

The $\mathcal{L}_1$-controller instead acts against this lowered control effectiveness to maintain the desired dynamics and sends a control signal to compensate for the input gain deficiency.

At some point, this has the same effect as high gain feedback and vehicle oscillations can occur. A changing input gain can be more harmful in terms of robustness than a changed damping of the plant. The $\mathcal{L}_1$-controller may in that sense be seen stricter than a PID-controller. Some of these ideas are also presented in [1].

Discussion of Potential Solutions

If facing robustness issues with one configuration of desired dynamics, solutions for adjusting those can be:

- Parameter scheduling: Scheduling variables are e.g. rotormast torque, RPM, airspeed, position of the collective lever (or actuator), altitude, side slip angle, angle of attack.

- Online parameter identification by

    - Regression for $\hat{\sigma}(t)$ into $\hat{f}(x(t))$ und $g(\hat{\omega}(t)b)$, i.e. splitting $\hat{\sigma}(t)$ into $b(\hat{\omega}(t) - \omega_0)u(t) + \hat{k}x(t)$.

    - Direct identification as in MRAC with a gradient-based adaptive law, if the condition of persistent excitation can be satisfied.

  See for example [27], where a neural network enhances the $\mathcal{L}_1$-controller.
  Results from parameter identification are usually sent to an online computation of an optimal controller or of a dynamic inversion controller.

- A sophisticated time varying state space model for the predictor with u, v, w (velocities in the three axes of the body-fixed frame).

- Replacing an online learning algorithm with an offline learning one. This corresponds to the general case of system identification with more options, e.g. nonlinear and unsteady dynamics. This could be done e.g. by:

    - Parameter identification creating charts for interpolation (look-up tables)

    - Training a neural network

  This is similar to the scheduling option but can take additional dynamics into account, i.e. not only changing the parameters a and b (example for first order plant), but also adding nonlinear terms that are canceled by the control signal.

- A sophisticated predictor based on physical knowledge in the plant, in particular energy models. Short term dynamics however are hardly improvable with this approach

– transient behavior in a barrel role can barely be modeled with it. The same applies to sideslip or changes in mass and other unmeasurable (or difficult to measure) conditions.

For online identification of the input gain, certain conditions have to apply and thus a certification in civil aerospace is unlikely. These are conditions of persistent excitation itself or similar concepts, that all have a common basis with offline system identification, namely the demand for sufficient information content in the input and output signals.

Besides the danger of statistical learning to lead to a non-robust solution, it is considered slow. A faster adaptation may be desirable. Ideally, an adaptation rate that is much faster and such that it is decoupled from the control signal regarding frequencies is sought.

Consequently, it is reasonable to gain schedule the parameters of the desired dynamics (predictor of the $\mathcal{L}_1$-controller), i.e. to 'scale' it, even if the baseline controller is not scheduled, cf. [1]. Scheduling is a commonly certified practice. Helicopter flight dynamics change significantly over the flight envelope. The best alternative to having one and the same behavior over the entire flight envelope is considered to 'scale' the dynamics. In case of a first order desired transfer function from command input to angular rate output, this means different damping and input gains, but still first order systems.

Only if crucial scheduling variables are not measurable or reconstructable with sufficient accuracy, at least one self-adapting parameter seems desirable.

Note that a continuous controller formulation with gradient based adaptive laws (based on Lyapunov design) is generally not suitable to parameter identification – especially identification of the plant's input gain – due to the general lack of required conditions. The adapted parameters thus cannot be considered a reliable estimation.

Chapter 3

Design of the Input Channel to the Predictor

This chapter addresses the generation of an input signal to the predictor – the predictor's analog to the total plant input. Solutions are shown for input channel modules that can be reused independently of the predictor form, but are an essential element in the tuning process of the $\mathcal{L}_1$-controller. Their design is shaped by modeled input time delays, saturation, dynamic elements and the choice of the inclusion or exclusion of trim and decoupling signals. Although mainly referring to augmentation architectures, the modules are straightforward to apply to standalone $\mathcal{L}_1$-controllers as the baseline control signals just have to be ignored.

The input modules are reusable across various helicopter types. If the decision has been made on a specific structure for every helicopter type, only saturation limits and the modeled amount of time delays (or actuator dynamics if necessary) need to be adjusted to the helicopter in question.

3.1 Input Signal Merging

The predictor input is based on the total plant input. Modifications can be made, however, e.g. by hedging certain signals from the predictor.

Digital inner loop commands may stem from an autopilot or a cockpit inceptor (e.g. sidestick for the pilot). These commands may enter the predictor of a standalone $\mathcal{L}_1$-controller directly. In augmentation, they are sent to the baseline controller, but not directly to the $\mathcal{L}_1$-controller. With the predictor however receiving the total actuator command vector (or actuator output vector), the baseline controller signal and thus the inner loop *commands* are included. Independent of the origin, this term is denoted as $r(t) \leq \|r\|_{\mathcal{L}_\infty} = \bar{r}$ in the predictor (sometimes contained in a $u_{BSL}(t)$). It is also possibly a mechanical signal that is serially augmented by a controller.

The command $r(t)$ is sent through a command filter $C_r(s)$. The filter roles vary: If the command is included in the baseline control signal, $C_r(s)$ *describes* the filtering structure of the existing baseline controller. If the command is directly sent to the predictor, it *introduces* a command shaping element.

In the latter case, if a greater bandwidth (less phase lag) in the command is desired than in the feedback loop, $C_r(s)$ can be chosen as $C_r(s) = C(s)^{-1}C_c(s)$, where $C_c(s)$ is a transfer function such that $reldeg(C_r(s)) = 0$, $reldeg(C_c(s)) = reldeg(C(s))$ and $C_c(s)$ has higher bandwidth than $C(s)$.

In augmentation, the filtering structure must not affect the baseline controller signal to preserve its structural integrity ($C_r(s)$ describes the already existing filters of the baseline controller); otherwise performance losses would occur.

For a standalone $\mathcal{L}_1$-controller, the command needs to be adjusted by some feedforward gain k_g for tracking the command with DC-gain one. In augmentation, the desired DC-gain is defined by the combination of the baseline controller with the predictor.

In some cases, the total plant input contains trim and decoupling functions, e.g. in augmentation or if a trim module is active.

Fact 1: Any predictor is interpreted as a description of dynamics around a trim point, unless changing trim conditions are modeled explicitly (similar to a constant disturbance).

Fact 2: A helicopter shows highly coupled responses on every axis (pitch, roll, yaw, heave), no matter where the original and potential only input is applied. Given that cross-couplings are hard to be identified, the question arises whether these can be modeled sufficiently in the predictor.

These facts guide to the need for dealing with trim and decoupling functions in the input channel to the predictor while complying with actual saturation effects. The latter is challenging as saturation is a nonlinear operation. Conditions for proper hedging are given in Appendix G.

3.2   Trim

The predictor describes dynamics around a trim point. In chapter 2.4, different kinds of trim are described. Attitude trim is part of the command and thus not of concern. Any other trim signal contained in the input channel (actuators!) is to be dealt with. Hence, the trim signal included in the input channel of the plant may be hedged from the predictor.

This fact implies a limitation to the application of predictor based controllers: In case of the input signal containing a trim signal, the numerical value of the trim signal has to be available – an issue that may guide to measurements of the actuator position or the application of fly-by-wire systems – whether for full or partial authority systems.

3.3   Decoupling of Cross-Couplings

It is chosen for a primary (i.e. preferable) architecture in this thesis:
a) for augmentation: To hedge decoupling functions of the baseline controller from the predictor (see Approach 5 in Appendix B), and
b) for the standalone $\mathcal{L}_1$-controller: To neglect knowledge about couplings (see Approach 4 in Appendix B).

The rationale behind the choices for the primary architecture is the strength of Approach 5 to be experimentally designed traditionally in frequency domain. It is therefore superior to a model approach in most cases as modeling cross-couplings by knowledge is accompanied by high design effort, if possible at all with sufficient accuracy. Furthermore, the command decoupling function has feedforward character and is therefore faster than any feedback controller. Only portions that cannot be canceled in the closed-loop plant are to be taken care of by the feedback controllers. If the decoupling signal is hedged from the predictor, a proper saturation of the input channel is guaranteed by proofing that the decoupling signal alone lies within saturation limits. A solution with its sufficient conditions is given in Appendix G.
The strength of Approach 4 may be its simplicity combined with the benefits of the $\mathcal{L}_1$-controller.

Migrating to a standalone $\mathcal{L}_1$-controller is considered preferable to any modification of the baseline controller in augmentation.

3.4   Structure

With alternatives shown in Appendix B, trim and decoupling signals (if existent) are hedged from the predictor in the primary architecture for reasons explained above.

The objective is to extract a signal from the total input vector of the plant such that it has the same basic information content, appropriate however for the predictor instead of the plant.

It is not useful to saturate the software input channel with tighter saturation bounds than the physical ones (in the plant input channel) as this may introduce distortions and harm the predictor's purpose in addition to the shrinkage of input authority. The structures are explained next.

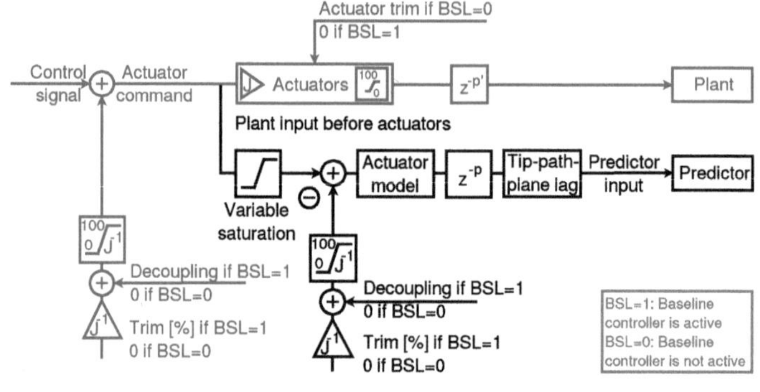

FIGURE 3.1: Predictor input channel – plant input signal *before* actuators

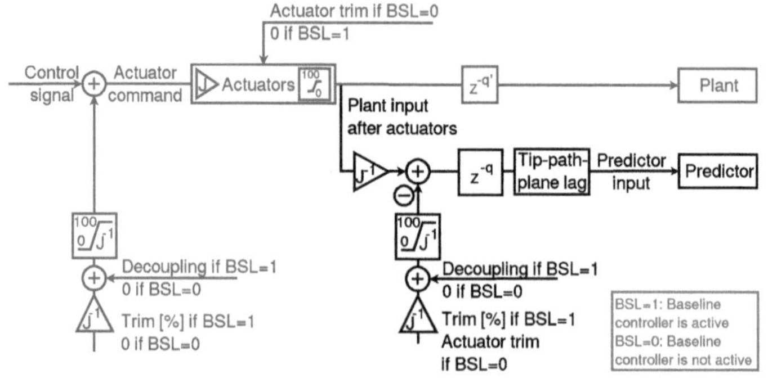

FIGURE 3.2: Predictor input channel – plant input signal *after* actuators

Explanations to Figure 3.1 and Figure 3.2:

- Both, Figure 3.1 and Figure 3.2 are possible structures for generating a valid input signal for a predictor. The difference between the options is how the total command vector is obtained – as total command vector *for* the actuators (e.g. a digital signal as sum of all controller commands) or the output *of* the actuator.

- Figure 3.1 and Figure 3.2 are intended for predictor dynamics that are valid (e.g. by system identification if a nominal model is desired) for an input signal, which is considered *before* the actuators (i.e. the actuators are virtually part of the plant dynamics), i.e. before the gain J is applied inside the actuators.

- The modeled time delays z^{-p} and z^{-q}, expressed in multiples of the computational time step, differ by the delay introduced inside the actuators. A more detailed explanation to time delays and the "tip-path plane lag" block for replacing some delay with a transfer function is given in Section 3.5.

- The gain J is part of the actuators and maps degree of blade angle to actuator position in percent. The gain J determines the saturation of 0...100% – also an actuator-inherent value. See Appendix O.4 for a more detailed explanation about actuators.

- A J^{-1} in the saturation block means that the saturation bounds are multiplied by J^{-1}.

- Both options from Figure 3.1 and Figure 3.2 have the principle in common to first saturate the sum of all control signals, i.e. the total input to the plant and after that to subtract other saturated signals, trim and decoupling signals. Note that this saturation architecture is different to what is shown in Figure 2.16. Conditions and a proof for this enhanced saturation structure are shown in Appendix G.

- The variable saturation of Figure 3.1 is specified in Figure 3.3.

- Figure 3.3 is based on the fact that $0 \leq J \cdot cmd + trim \leq 100$ where cmd is the actuator command vector from controllers that do not add trim and decoupling signals, i.e. in case a baseline controller does not exist.

- The structure for adding trim and decoupling can equivalently be implemented in different forms as shown in Figure 3.4.

- If the actuator dynamics are sufficiently fast, the actuator model can be simplified to the respective saturation and an equivalent time delay – or a lower order transfer

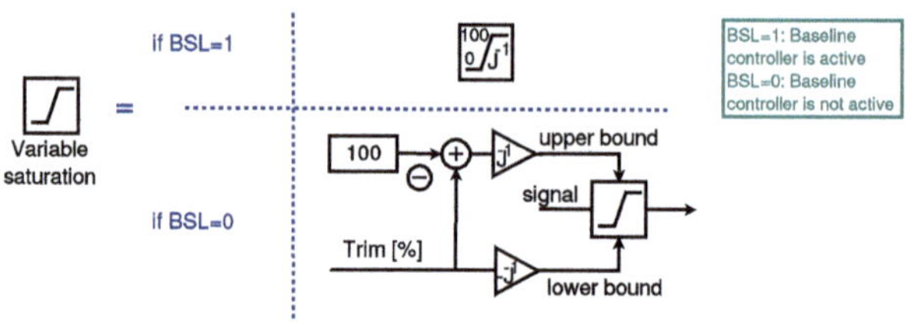

FIGURE 3.3: Predictor input channel – variable saturation for signal before actuators

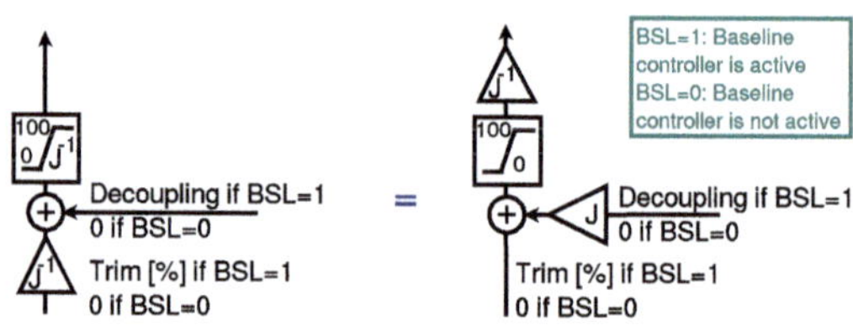

FIGURE 3.4: Predictor input channel – alternative trim structure

function. The higher the actuator bandwidth, the smaller the deviation introduced with this simplification. Despite a loss of precision, software complexity can be reduced.

- In case of an active baseline controller being present, trim and decoupling functions are considered part of the baseline controller. In absence of a baseline controller, the original trim position of the actuators (static trim) is considered part of the actuator, i.e. it is a signal that is not visible when obtaining the control signal prior to the actuators. This explains the difference between Figure 3.1 and Figure 3.2 in the subtraction of trim in dependence of the connection of a baseline controller.

If possible, the signal after the actuators is to be preferred to the one before – as long as reliable information about the current actuator position can be provided. With this, a solution for "adapting" the controller upon actuator anomalies is covered automatically. It simplifies the design by requiring less system knowledge and takes non-nominal behavior of the actuator into account – a perfect solution for maximizing robustness in case of actuator failures. If for instance the actuator jammed and saturated earlier in regard to its position or rate limits, this degradation would be accepted as the currently available input to the plant by the predictor – in contrast to fighting it. Beyond dealing with failures, a higher precision is achieved in general as parts of the modeling efforts are replaced with measurements.

The closer the definition of the "input signal" to the rotor in the entire input channel, the higher the safety regarding actuator failures and the higher the quality of the prediction value due to minimizing the total amount of sources of dynamical effects and with it the amount of unmodeled dynamics. This stands against an additional safety-critical sensor at the actuator.

It shows how the term "input" is in general vague and it is to be defined at which point of the input channel the "input" is considered.

3.5 Modeling of Response Lags

Several effects can be considered for estimating their contribution to a total time delay of the input. This may include material elasticity of mechanical parts, actuator delay when "input" means "prior to the actuators", some delay due to the phase lag of the rotor (the response in the flapping motion due to a change in the blades' AoA), the inertia of the surrounding air and the airflow through the rotor. In addition, slight non-minimum phase characteristics of the rotor can be dealt with conveniently by adding more modeled time delay.

Time delays in the sensor, buses and processors, e.g. in the IMU, (all *not* in the input channel) are merged into the input channel. Simply copying the known sensor latency into the input channel of the predictor may be a good point to start with for tuning but is not necessarily identical to the optimum amount of time delay to be added due to sensor delays.

Actuator delays are hardware inherent parameters. Effects of material elasticity are usually small; however, tail rotor delay is noticeably higher due to the longer mechanical signal path [28].

For estimating the delays in the flapping response of the main rotor, the phase lag is considered. For instance, with assumed exemplary data of the rotor, radius $r = 5\ m$, phase delay due to flapping: $\Delta\Phi = 70\ deg$, and blade tip velocity: $v_{tip} = 215\ m/s$, a first guess of this time delay is modeled by $\tau_1 = \frac{2r\pi}{v_{tip}} \cdot \frac{\Delta\phi}{360\ deg} \approx 0.028\ s$.

However, the flapping motion builds up continuously rather than in form of a step. Hence, in contrast to modeling a simple time delay, a dynamic transfer function can replace some of it, cf. Figure 3.5. This may be closer to reality (although not significantly) and provides the possibility to take non-minimum phase behavior more accurately into account.

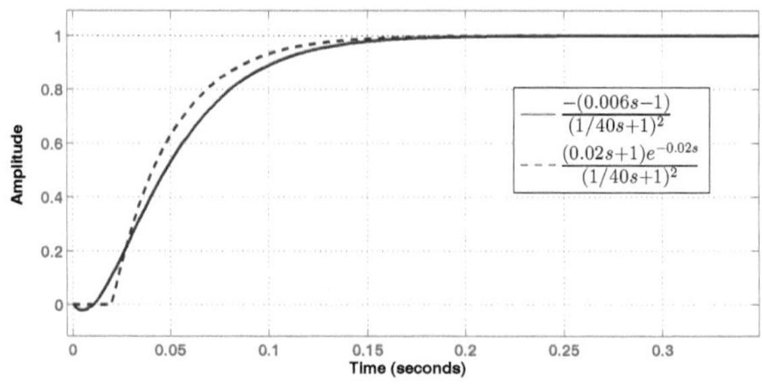

FIGURE 3.5: Predictor input channel – step responses of different models for the tip-path plane lag

It provides an estimation for the time delay to be modeled. Eventually however, tuning aims at closed-loop system bandwidth and the trade-off between performance and robustness. If the amount of modeled time delay is too high, the closed-loop system bandwidth suffers and the higher the chance of facing unwanted overshoots in the system's response.

This degree of freedom for tuning the performance comes with effects on the time delay margin. There is an optimum of the amount of modeled time delay in regard to the time delay margin. In some simulations it can be observed that adding slightly more time delay than what is set up in the plant's input channel helps to increase robustness, i.e. the time delay margin. More modeled time delay can be observed to decrease the closed-loop system bandwidth.

Chapter 4

$\mathcal{L}_1$-Control for Pitch, Roll, Yaw

This chapter presents $\mathcal{L}_1$-controllers in state feedback and in output feedback for pitch, roll, and yaw – solely geared to the primary architecture. Alternatives are shown in the appendix. Conceptually related for example are [29], [30], or [31].

In state feedback it refers to both, augmentation and standalone $\mathcal{L}_1$-control. Although architecturally suitable for augmentation, the output feedback controller is applied only to standalone $\mathcal{L}_1$-control. Augmentation in output feedback is considered not advisable herein and its benefits moot due to its robustness issues and a required complexity exceeding the one of legacy control laws by far. Furthermore, the constraints on the desired transfer function in the structures shown $(reldeg(M(s)) = 1)$ cumber the model fitting in identified frequency responses.

The augmenting and the standalone $\mathcal{L}_1$-controller share the same structure. In augmentation, the parameters of the desired dynamics are chosen as a best guess of nominal helicopter dynamics, whereas in standalone control more freedom in choosing the parameters is available – be it faster or slower than the nominal dynamics. Some ideas of augmentation are shown in [1].

4.1 In General

Figure 4.1 depicts a general structure. The block "Model. in.-dyn" stands for "modeled input dynamics" and is required only in some cases.

The predictor provides an estimation of the closed loop states $\hat{x}(t)$ and output $\hat{y}(t)$, respectively with a form of uncertainty estimation, namely $\hat{\sigma}(iT)$ from the processed plant input $u(t)$.

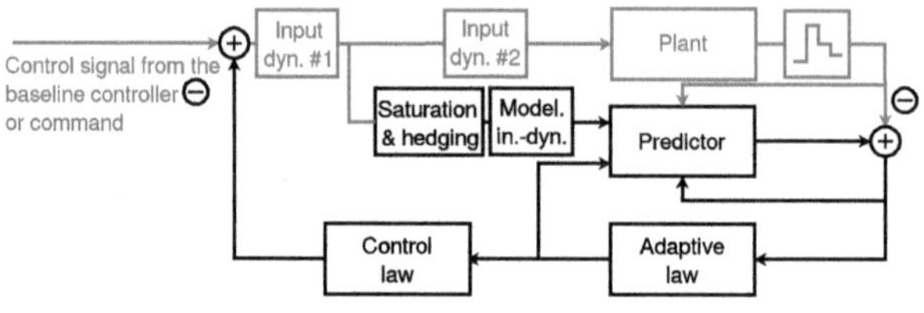

FIGURE 4.1: Implemented $\mathcal{L}_1$-controller

The adaptive law provides $\hat{\sigma}(iT)$. For the piece-wise constant adaptive law, $\hat{\sigma}(iT)$ contains the time step T and is thus not a real estimation in the sense of online identification, in addition to all uncertainties and disturbances being lumped into one single variable.

The control law sends a low-pass filtered version of $\hat{\sigma}(iT)$ to plant and predictor, thus canceling uncertainties in a low frequency range and adding high frequency uncertainties to the desired dynamics, hence the transition from desired dynamics to a predictor.

The input module to the predictor is described in Chapter 3. It accounts for hedging trim and decoupling functions if existent and provides primarily input saturation and a time delay similar to the one in the physical system.

4.2 State Feedback

The predictor states are (predicted) compensated angular rates in roll, pitch and yaw $\hat{\omega}_{comp}(t) = [\hat{p}(t), \hat{q}_{comp}(t), \hat{r}_{comp}(t)]$ upon the input (either before or after actuators). Compensated means that turn rates in pitch and yaw are subtracted from the measured rates in these axes – the roll turn rate is ignored as it is usually small unless performing aerobatics, where turn compensation is not active.

Three decoupled SISO predictors are selected for the primary architecture. With $x_k \in \{p, q_{comp}, r_{comp}\}$ (see Chapter 2.4 regarding *comp*ensated turn rates), the predictor for each axis k is defined as:

$$\dot{\hat{x}}_k(t) = a_k x_k(t) + b_k(u_k(t) + \hat{\sigma}_k(t)) + a_{SP}\tilde{x}_k(t); \quad \hat{x}_k(0) = x_k(0) + \nu_k(0) \qquad (4.1)$$

where for every k: $a < 0$ and b define the stable desired dynamics, $a_{SP} < 0$ defines the stable error dynamics ($a_{SP,k}$ would also be possible), and the prediction error is defined as $\tilde{x}_k(t) = \hat{x}_k(t) - x_k(t) - \nu_k(t)$. The predictor states are initialized with the measured states: $\hat{x}_k(0) = x_k(0) + \nu_k(0)$, where $\nu_k(t)$ is the measurement noise on the respective axis.

The control law is defined as:

$$u_k(s) = C_{r,k}(s)k_{g,k}r_k(s) - C_k(s)\hat{\sigma}_k(s) \tag{4.2}$$

where for every k: $C(s)$ is a strictly proper and stable low-pass filter with DC-gain one. It is initialized with zero. The gain $k_g = -(a^{-1}b)^{-1}$ ensures the correct DC-gain in tracking.

The adaptive law is defined as:

$$\hat{\sigma}_k(t) = \hat{\sigma}_k(iT) = \frac{a_{SP}e^{a_{SP}T}}{b_k\left(1 - e^{a_{SP}T}\right)}\tilde{x}_k(iT); \quad i = 0, 1, 2, ... \tag{4.3}$$

Alternatively, a recursive adaptive law can be applied:

$$\hat{\sigma}_k(iT) = \frac{a_{SP}e^{a_{SP}T}}{b_k\left(1 - e^{a_{SP}T}\right)}\tilde{x}_k(iT) + \frac{-a_{SP}}{b_k\left(1 - e^{a_{SP}T}\right)}h_k(iT); \tag{4.4}$$

where

$$h_k(iT) = -\tilde{x}_k(iT) + h_k((i-1)T); \quad h_k(0) = 0; \; i = 1, 2, ... \tag{4.5}$$

A recursive term to the otherwise proportional gain is added for introducing integral behavior. This provides a significant decrease of $|\tilde{x}(t)|$ which otherwise could only be achieved with much smaller time steps T. It can hurt the time delay margin however, see section 2.5.2 and [19].

With the hedging of decoupling functions, the controller can be implemented with three independent SISO controller modules instead of a decoupled (3×3) MIMO system. Then it is safe to compute $\hat{\sigma}_k(t)$ online when scheduling a_k and b_k since matrix inversion is replaced with scalar division.

Possible Extensions:

Additional predictor dynamics can be added: The rotor system is not a rigid body – the blades themselves are far from being rigid and are free to flap and lag per revolution. Thus, the rotor is not a gyroscope but shows similar effects. These effects can be added to the predictor as additional term that requires its counterpart in the baseline controller – a term that cancels exactly this modeled coupling term. This method is described in Approach 3 in Appendix B. See Appendix K for inertial couplings including an optional term modeling gyroscopic effects of the rotor.

Baseline controller states can be included in the predictor. This method sees the $\mathcal{L}_1$-controller as the outer loop of an inner loop baseline controller. The contrary point of view is justified, too. As long as unmatched uncertainties are not considered, both options can be tried freely. Appendix M describes this option for the baseline controller shown.

Parameter scheduling can be applied. The general control goal is that the plant tracks the desired dynamics. This is true for the shape (e.g. first order response) of desired behavior when scheduling desired dynamics. At the same time, the controller "follows" approximately the plant regarding "aggressiveness" of the fixed shape. For reasons discussed in section 2.7, it can be necessary to schedule the predictor parameters while the baseline controller is not scheduled. This is the case for instance in augmentation. Scheduling applies to a and b, and with it the adaptive law. It is done in this thesis mostly over airspeed. Other possible scheduling variables would be rotor RPM, which however in conventional helicopters remains about constant. Scheduling serves the purpose of acknowledging changing dynamics to be nominal. Autorotation for instance is not considered a failure case, but part of the nominal flight envelope. In this thesis, the scheduling is extended to a mild parameter change for very low collective lever positions. Scheduling by interpolation and an analytic function are shown in Figure O.3 on page 219.

Adjusting the filter bandwidth online. Besides *scheduling*, theory exists for *adapting* the filter bandwidth. See [32].

4.3   Output Feedback

This output feedback architecture can be integrated with the same function as the state feedback controller. That is, the output predictor uses the same scalar input-output pair as the state predictor.

This primary architecture is confined to the application of the recursive piece-wise constant adaptive law (cf. Equation (2.22) on page 34). While in state feedback it is considered optional, in output feedback it shows to be necessary in this thesis' application as a_{SP} is not existent in this form.

Define an output predictor such that with $\hat{\sigma}(t) \equiv 0$ it represents the desired dynamics. Everything not to be subsumed under the desired behavior is lumped into a vector $\hat{\sigma}(t)$.

With $y_k \in \{p, q_{comp}, r_{comp}\}$ and $x_k(t) \in \mathbb{R}^{n_k}$ being some internal states in every axis one has:

$$\dot{\hat{x}}_k(t) = A_k \hat{x}_k(t) + b_k u_k(t) + \hat{\sigma}_k(t), \quad \hat{x}_k(0) = \hat{x}_{0,k}; \quad \hat{y}_k(t) = c_k^T \hat{x}_k(t) \tag{4.6}$$

where A, b, c^T are a minimal state space realization of the desired dynamics $M(s)$.

It is initialized with some $\hat{x}_{0,k}$ such that $c_k^T \hat{x}_{0,k} = \hat{y}_k(0)$ and $\|\hat{x}_{0,k}\|_\infty \leq \rho_{\hat{x}_0}$.

The control law is defined as:

$$u_k(s) = C_{r,k}(s)r_k(s) - C_k(s)\frac{M_{um,k}(s)}{M_k(s)}\hat{\sigma}_k(s) \tag{4.7}$$

where $C_{r,k}(s)$ and $C_k(s)$ are strictly proper and stable low-pass filters that are initialized with zero. The desired transfer function for every k is: $M(s) = c^T(s\mathbb{I} - A)^{-1}b$ and $M_{um}(s) = c^T(s\mathbb{I} - A)^{-1}$. Implementing the filtering structure as state space model eases parameter scheduling regarding desired dynamics.

The adaptive law is defined as:

$$\hat{\sigma}_k(t) = \hat{\sigma}_k(iT) = -\Phi_k^{-1}(T)\Upsilon_k(T)1_1\tilde{y}_k(iT) + \Phi_k^{-1}(T)1_1h_k(iT), \quad i = 0, 1, 2, \dots \tag{4.8}$$

where for every k, $1_1 = [1, 0, ..., 0] \in \mathbb{R}^n$, and
$h(iT) = -\tilde{y}(iT) + h((i-1)T), \quad h(0) = 0, \ i = 1, 2, \dots.$

$\Lambda = \begin{bmatrix} c^T \\ D\sqrt{P} \end{bmatrix}$, with D such that: $D\left(c^T\sqrt{P}^{-1}\right)^T = 0$

and P such that $A^T P + PA = -Q$ and Q as design element such that $Q = Q^T > 0$.
$\Phi(T) = (-\Lambda A\Lambda^{-1})^{-1}\left(\mathbb{I} - e^{\Lambda A\Lambda^{-1}T}\right)\Lambda$, and $\Upsilon(T) = e^{\Lambda A\Lambda^{-1}T}$.

In augmentation, $M(s)$ must only show what the baseline accounts for. Other constraints are $reldeg(M(s)) = 1$, $M(s)$ must be chosen stable and minimum phase, zeros and poles of $M(s)$ must be sufficiently far away from zero to avoid lightly damped poles and zeros, which otherwise may evoke peaking phenomenon. Note that $M(s)$ is inverted in the control law. Furthermore, $C(s)$ and $M(s)$ must be chosen for $H(s)$ to be stable (see Appendix F) and should not be more complicated than necessary to avoid overly high CPU loads.

Initialization Procedure:
As a full set of measurements of internal states is not available for a complete controller initialization as in state feedback, the controller operation is initiated in broken-loop mode. This allows for the internal dynamics to converge to reasonable values and the transient error in the input signal to settle. The idea is first shown in [33].

The steps of the starting procedure are defined as follows:

1. Initialization: Internal states $\hat{x}_k(0)$ are initialized such that $c_k^T \hat{x}_k(0) = \hat{y}_k(0) = y_k(0) + \nu_k(0)$, i.e. $\tilde{y}_k(0) = 0$.

2. To avoid undesired transients in the control signal originating from $\tilde{x}(0) = \tilde{x}_0$, the controller is run in broken-loop mode for a certain time span. How to tune this time span can be concluded from the time constants of the later shown Equations (F.107) and (F.108) where $\frac{C(s)}{M(s)} y_{in}(s)$ decays exponentially fast.

3. The loop is closed and the controller operates in normal mode.

Figure 4.2 shows where the loop is disconnected to operate in broken-loop mode.

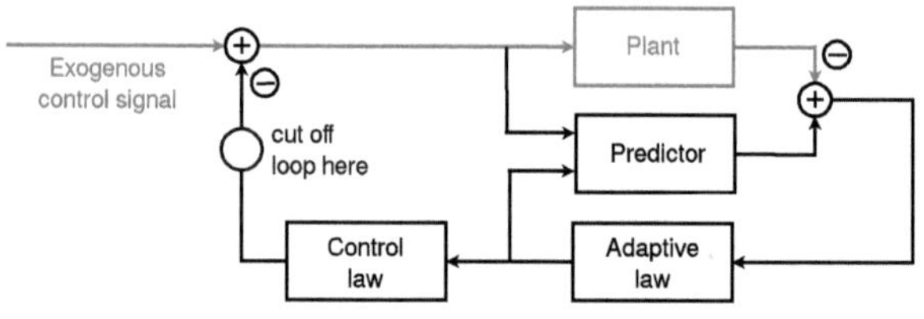

FIGURE 4.2: Broken-loop $\mathcal{L}_1$-controller

Initializing shortly after takeoff can be practically reduced to setting the internal states to zero as the angular rates are close to zero.

Possible Extensions:
Additional predictor dynamics, parameter scheduling, and adjusting the filter bandwidth online from section 4.2 apply equally to output feedback. It is difficult however to incorporate the baseline states since this architecture in general works with unknown internal states.

4.4   Outer Loops for the Standalone $\mathcal{L}_1$-Controller

This section shows simple outer loops with proportional errors. The $\mathcal{L}_1$-controller resembles a PI-controller (cf. [13] or [15]). Thus, if well tuned, applied to the rates only it approximates an RCAH system. However, using available attitude sensor information in addition may be helpful.

Enhanced structures can also be found for example in [10] or [28].

Figures 4.3, 4.4, and 4.5 illustrate outer loops for rate command in pitch, roll, yaw, and attitude command in pitch and roll. Figure 4.6 shows a velocity command system for pitch and roll.

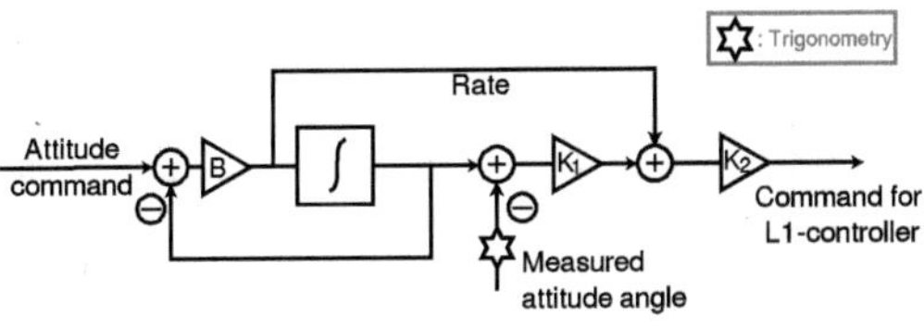

FIGURE 4.3: Nonadaptive outer loop in **pitch and roll** for adaptive inner loop – attitude command

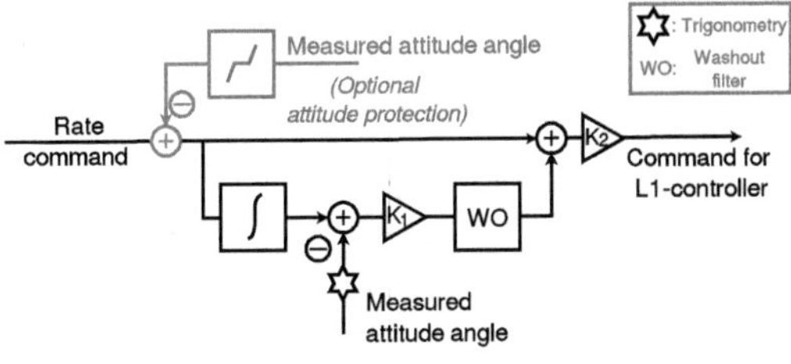

FIGURE 4.4: Nonadaptive outer loop in **pitch and roll** for adaptive inner loop – rate command

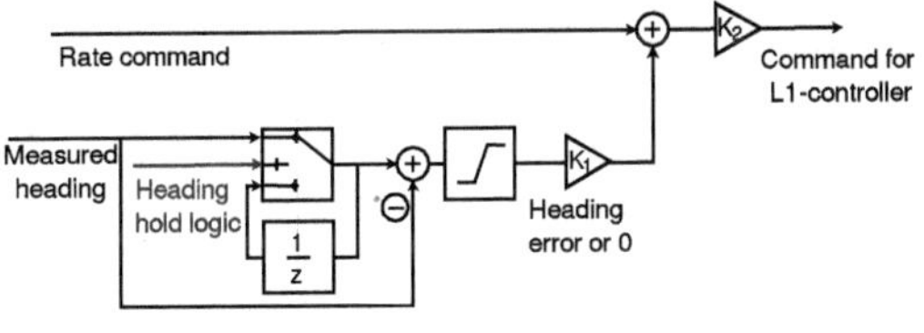

FIGURE 4.5: Nonadaptive outer loop in **yaw** for adaptive inner loop

In rate command of pitch and roll, a high-pass (washout) filter is introduced to the attitude channel for avoiding overly strong reactions on disturbances affecting the attitudes. In rate command mode it is expected that a gust moves the vehicle to a different attitude, without the original attitude being recovered immediately. Yet the attitude information is partially used.

- The gain K_1 is a feedback gain and is thus tuned according to a trade-off performance and robustness.

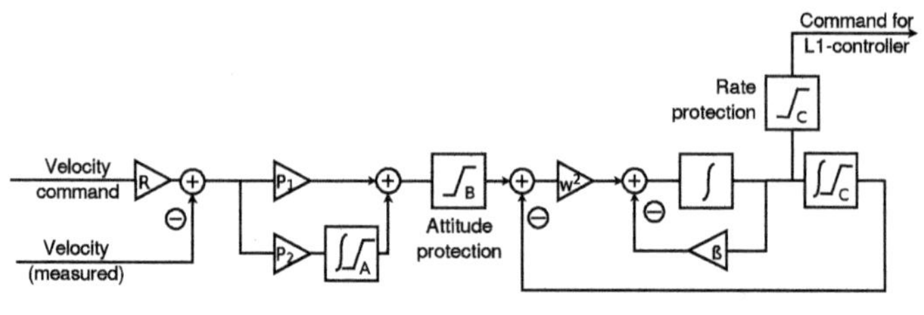

FIGURE 4.6: Nonadaptive outer loop in **pitch and roll** for adaptive inner loop – velocity command

- The gain K_2 determines the DC-gain of the desired dynamics. In state and output feedback it is computed as $(-a^{-1}b)^{-1}$ and 1, respectively, so that the controller tracks this command with DC-gain one.

Rate and attitude command refer to the body-fixed frame. Instead of commanding Euler angles, an "integrated body-fixed rate" defines the commanded attitude. With the feedback stating Euler angles, important elements of the system transformation are added for the pitch axis as shown in the Figures 4.3 and 4.3. Transformations in the roll and yaw axis are neglected.

Attitude protection is provided naturally for ACAH systems by motion limits of the inceptor and the autopilot, respectively. In RCAH systems it is optional and switched off during aerobatics as every other attitude module. Attitude protection is most important in roll for avoiding dangerous conditions of the spiral mode.

Note that the outer loops are nonadaptive and without integrators. Similar to two cascaded integrators, another $\mathcal{L}_1$-controller in the outer loop eliminates unique adaptation since the relation from rates to angles is uncertainty free.
An exception is the velocity command system shown in Figure 4.6. In that case, an integrator is required for avoiding position drift.

Chapter 5

$\mathcal{L}_1$-Control for Vertical Speed in Hover

This chapter presents a vertical speed $\mathcal{L}_1$-controller for imposing artificial stable first order behavior and disturbance rejection on the heave motion, confined to hover and low speed regimes. Only a standalone controller in state feedback is presented here. This seems legitimate as internal dynamics are expected to be negligibly small. A seamless control initialization is sought. Ideally the pilot does not notice controller activation and support. The collective lever position zone for $\dot{h}_{cmd}(t) \equiv 0$ however can be shown in a visual cue. This zone emerges from the initialization position surrounded by a dead zone. It is aimed at a robust design so that drift can be counteracted to regardless of mass changes.

The content of this chapter is introduced in [1] and published in [2] by the author; this chapter shares figures, equations and text with the aforementioned papers.

5.1 Background

In hover and low speed regimes, requirements for precision flight are usually the most demanding ones. Attaching or jettisoning a load, having personnel rappel, hiding behind objects – precision is not only a demand for the mission, hovering near ground and objects also arouses safety issues. Especially divided attention and degraded visual conditions add to work load and safety concerns. Thus, a stability augmentation system is introduced here.

For controlling the heave motion, robustness against mass changes is challenging. Expected mass changes with factor 2 between empty mass and MTOM (maximum takeoff mass) present a huge parametric uncertainty, unless a mass estimation is applied on which desired dynamics are scheduled upon. Mass estimators however are usually inept at sudden mass changes as when jettisoning an external load. Thus, the controller is designed to be robust against mass changes.

Sensor signals for vertical speed stem primarily from the IMU (inertial measurement unit), but can be supported by filtered derivatives of the radar altimeter signals in hover near ground.

5.2 Controller Core

The vertical speed is the only state of consideration, $x(t) \equiv \dot{h}(t)$.

The state predictor is scalar:

$$\dot{\hat{x}}(t) = ax(t) + b(u(t) + \hat{\sigma}(t)) + a_{SP}\tilde{x}(t); \quad \hat{x}(0) = \dot{h}(0) + \nu_h(0) \tag{5.1}$$

where a and b state desired dynamics, a_{SP} shapes the error dynamics, $\dot{h}(t)$ is the vertical speed and $\nu_h(t)$ is the sensor noise. The input $u(t)$ is described in detail in the next section of this chapter.

The predictor, i.e. the integrator of its differential equation, gets initialized with the actual $\dot{h}(0)$.

The adaptive law is defined as:

$$\hat{\sigma}(t) = \hat{\sigma}(iT) = \frac{a_{SP}e^{a_{SP}T}}{b(1 - e^{a_{SP}T})}\tilde{x}(iT); \quad i = 0, 1, 2, ... \tag{5.2}$$

The control law is defined as:

$$u(s) = C_r(s)r(s) - C(s)\hat{\sigma}(s) \tag{5.3}$$

where $C_r(s)$ and $C(s)$ are strictly proper and stable low-pass filters, initialized with zero.

Note that these equations state only the core; saturations and the entire structure is shown later.

Design of Desired Dynamics

The natural response of the vertical speed $\dot{h}(t)$ on a collective step input is shown – slightly simplified – in Figure 2.2. First order desired dynamics regarding "collective input $\leftrightarrow \dot{h}$" for hover hence are a realistic control objective. Any non-minimum phase behavior can be merged into the modeled input channel to the predictor.

In hover, the collective input generates a force (lift/gravity) on a mass (helicopter) – a one-dimensional mass-damping system.

The desired dynamics – defined by the predictor – are *not* designed conservatively, but to state agile behavior that is possible only with low masses. This is allowable due to two protection mechanisms:

1. The input $u(t)$ to the predictor is simultaneously saturated with the input offset from trim to the helicopter. Thus, the desired dynamics automatically limit the performance being asked for because of saturation.

2. In $\mathcal{L}_1$-control, high frequencies in the control channel are considered not possible to compensate for and are excluded from the desired dynamics.

These features protect against wind-up and ensure that the prediction error remains small with low and huge masses, without sacrificing agility in case of flight with low masses.

Avoiding Altitude Drift

A large mass change shifts the trim – a new offset the $\mathcal{L}_1$-controller has to compensate for. Large disturbances increase $|\tilde{x}(t)|$ which can propagate to the tracking error, hence introducing an offset in the desired vertical speed – drift occurs.

This mechanism is shown in Introduction 2.5.4 with some remedies. For convenience, possible solutions are repeated herein: Slow a_{SP}, a recursive adaptive law, an outer loop for direct altitude control, an exogenous trim signal, or shorter time steps T can be applied. Slow a_{SP} and the recursive adaptive law may show robustness issues, shorter T are rarely feasible for sufficient effects, and an outer loop is prone to introduce a reverse motion, e.g. when jettisoning a load. The primary architecture uses slow a_{SP}, see also Chapter 7.

Filter Bandwidth Trade-Off

When jettisoning a load, a quick counter-reaction in the heave motion is desirable. Thus, a large filter bandwidth is necessary. Robustness analysis can be conducted with high confidence since in the scalar problem the parameter space to be tested is very limited. Flying in different altitudes with different masses allows for a complete evaluation.

5.3 Command Signal Processing

This section deals with generating the vertical speed command from a collective lever input.

Overview

When activated, the command value is stored as trim. When active, the trim value is subtracted from the command such that the vertical speed is commanded around this trim position. The trim position, i.e. zero vertical speed command lies within a dead zone. The bounds of the dead zone are variable due to an added hysteresis. This provides a vertical speed command until the controller is switched off.

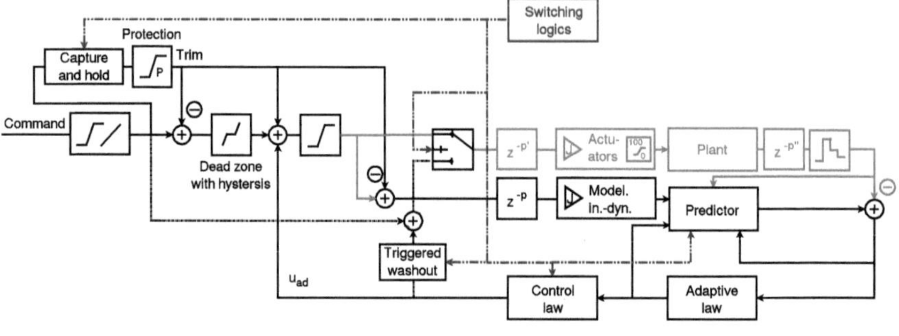

FIGURE 5.1: Architecture of the vertical speed controller

A detailed explanation of the elements in Figure 5.1 is given next.

Cockpit Inceptor

Assume a collective lever in fly-by-wire architecture. Its output is rate- and position-limited, where the rate limiter is generally faster in the initial response than a dynamic command model, but provides a command that is not fully trackable.

The collective lever provides a command for collective blade angles, a value roughly proportional to rotor thrust, but could also be used in fly-by-wire systems to command vertical speed for instance.

Activation Logics

A manual switch integrates or disintegrates this module. If on, the controller activates automatically shortly after takeoff or within a defined attitude and speed regime. If hover is not desired during takeoff, the collective lever is just pulled further upwards which is equivalent to command an altitude gain. Exceeding any of the speed or attitude thresholds, the controller is deactivated. The block "Switching logics" in Figure 5.1 refers to these functions.

Vertical Speed Command Generation

If activation is triggered, the position of the collective lever is saved as **trim** value and subtracted from the subsequent commands. This refers to the block "Capture and hold" in Figure 5.1. It applies within a certain region of the collective lever position, the trim is subject to smaller limits than the bounds of the inceptor signal, shown in the block "Protection". (E.g. if the inceptor command reaches from 0% to 100%, the protective saturation may reach from 15% to 85%.) This leaves some margin to the commanded vertical speed, so that altitude hold in hover does not coincide with a maximum collective lever position.

A **dead zone** wrapping the difference (zero at activation) of command and trim is set. This dead zone serves as margin for the collective lever position to avoid undesired commanded drift and for finding the equilibrium of $\dot{h}_{cmd}(t) \equiv 0$ more easily when turning back from a commanded nonzero vertical speed.

The dead-zone is complemented by a **hysteresis** function that ensures that if the collective lever is driven into the dead-zone and being held at the bound of activation, the dead-zone is not immediately left or affected by high frequency switching. The bounds from the protection block are chosen in accordance with the magnitude of the hysteresis and dead zone. This guarantees command authority for the pilot in both directions, descending and ascending, which otherwise may be concealed in the dead zone.

Saturation Architecture

The concepts explained in Chapter 3 apply: The sum of the trim and the command around it are saturated. Both summands are designed to not reach the saturation alone.

After the dead zone for the vertical speed command, the trim and the adaptive input are added. It follows the total plant input, yet not saturated. Saturating the total actuator command, it is safe to subtract the trim from this signal to the predictor under conditions shown in Appendix G. The result is the remaining effective signal in consideration of the absolute saturation. Thus, plant and predictor get the same input signals with the trim as the only difference.

The block "Model. in.-dyn." (modeled input dynamics) is optional and can replace some modeled time delay with actuator dynamics, and optionally rotor dynamics regarding mostly the tip-path plane lag. The actuator gain J however is always necessary.

5.4   Extensions and Enhancements

This section shows enhancements and variations that can be applied if desired.

Shift of Trim Point

When leaving the active regime, e.g. pushing nose down over the attitude threshold or gaining sufficient speed, the transition is hardly noticeable for the pilot as the collective command is very close to what it was when still active, that is trim plus possibly vertical speed command. This is not true if in hover an external load is attached or jettisoned. The trim point is then shifted which would result in an odd collective lever position for the new mass. Three solutions are proposed:

1) The adaptive control signal is sent through a washout filter when deactivated, thus rendering the signal change smooth instead of jumping.

2) An adjusting collective lever repositioning, meaning that the lever is moved slowly to the new trim position after jettisoning.

3) A fly-by-wire sidestick replacing the collective lever; it commands vertical speed in hover and something else in other conditions. Hands-off in hover means zero command.

Emergency Switch-Off

This controller is switched off in case of engine failure, i.e. at a threshold of power loss. With the controller counteracting any altitude loss, rotor RPM would be lost very quickly preventing a safe transition into autorotation. Note that in hover much more power is required than at medium forward speed.

Parameter Scheduling

Desired dynamics are determined by (a, b). The system matrix a describes the time constant, where b is linearly determining the DC-gain $\left(-\frac{b}{a}\right)$. The input gain b (control effectiveness) is mainly varying with helicopter mass and air density. In addition, it changes when in ground effect. Hence, if at all, it may be more important to schedule the DC-gain via b. With eschewing a mass estimator, scheduling may reduce to an altitude dependence regarding air density and ground effect. As the system is scalar, an either interpolated b or one from an analytic function is sent to the adaptive law, which can be computed online since matrix inversion is replaced with scalar division. Scheduling adjusts desired dynamics to the real situation and thus renders the controller more robust.

Outer Loop

While small values of $|a_{SP}|$ ensure satisfying disturbance suppression, the entire concept is still a rate controller, which lacks moving back to the original altitude after a disturbance. This may be preferred to recovering the original altitude as this is not very natural – except when in ground effect where the equilibrium may be at a certain height above ground due to increasing lift near ground. If necessary though, a proportional altitude error feedback, initialized simultaneously with the vertical speed controller can be added to the command. Integrators in the outer loop are prone to contribute to overshoots if not oscillations since the inner loop $\mathcal{L}_1$-controller provides approximate steady state accuracy as a PI-controller does, cf. Appendix E or F.

Limited Controller Authority

The adaptive input can be limited for either safety reasons or hardware limitations to solely *augment* on a hardware level.

5.5 Alternative Structure

The structure outlined in Figure 5.2 (simplified) poses an alternative to the one shown above. Its advantage is the removal of the dead-zone from the feedforward channel of the plant. Thus, as long as a numerical value for the direct command is known, the input channel to the plant can be mechanical with a serial controller support element.

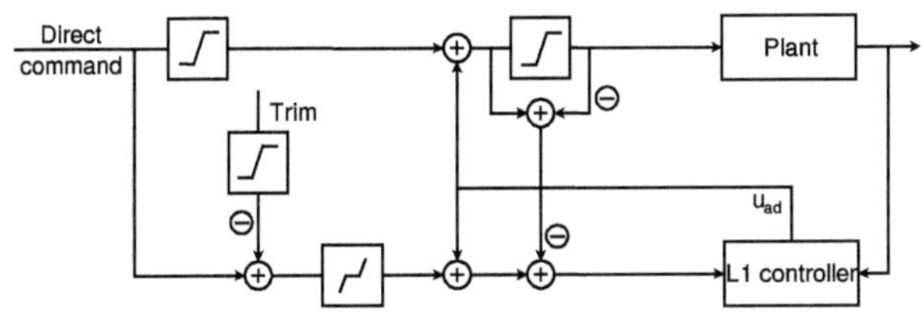

FIGURE 5.2: Alternative architecture of the vertical speed controller – simplified

The block "L1 controller" contains the input channel, i.e. the actuator gain, modeled time delays, optionally actuator dynamics and modeled dynamics of the rotor, e.g. the tip-path plane lag.

This architecture may be better fitting to traditional, non-fly-by-wire architectures. Its advantage is that even with loss of the serial actuators, a sufficiently strong pilot can still manually control the collective input.

Note that any direct command that lies within the dead zone is counteracted to by the serial actuator. This is the drawback of this architecture.

Chapter 6

Certification Strategy

This chapter suggests a strategy for the proposed $\mathcal{L}_1$-controllers that combines classic procedures with an argumentation adjusted to the novel approach. The strategy is confined to the top level, i.e. the algorithmic layer, however with respect to the connection of the software (code) layer. As mentioned in this chapter, the *software level* of the $\mathcal{L}_1$-controllers shown is very similar to the one of legacy controllers since the same elements as in the baseline controller are used.

6.1   Introduction

Flight control systems are safety-critical. Hence, they are a mandatory element of aircraft certification. The more authority the control system has in terms of input bandwidth and amplitude, the higher may be its criticality. With a full authority fly-by-wire system potentially causing an actuator hardover, failure can easily yield a catastrophic outcome. Only a sophisticated, deterministic design with formal and practical verification is acceptable. In addition, different types of controllers with comparable performance in nominal conditions may react differently to degraded dynamics, depending on the type of degradation. Targeting a new controller type to certification, an equivalent or higher level of safety is required.

Established certification procedures refer mainly to PI controllers. Neither helicopter dynamics nor the PI controllers themselves are linear due to switching and saturation. Linear performance and robustness metrics however are accepted due to the proximity of helicopter dynamics to linearity (at least locally) and the sufficient robustness against uncertainties. The latter is expressed by the gain and phase margins, accounting for sufficiently small gains and the distance to a phase shift of 180 *deg*. As a classic means of compliance these margins can support the certification process but are not mandatory. Moreover, current procedures heavily rely on testing. Taking up linear systems theory, the system is believed to operate well within the envelope if it is shown to operate well on the limits of the envelope. However, anything apart from a PI controller is considered non-standard. It is thus expected that these procedures toughen towards stricter and more preceding simulation tests, where the final flight tests are still an important element, especially to verify simulation results. The shortcoming of testing however is that it is rather a searching for errors than proofing the system correct [34].

As mentioned above, the key to certification is deterministic behavior, be it linear, nonlinear, or linear time varying. No control algorithm whatsoever however can be certified by formal mathematics alone. Relying on abstract mathematical results is practiced often, but cannot be accepted. A simple asymptotic convergence proof or signal that is proved to be bounded is rather meaningless for flight safety without transient behavior and rate of convergence combined with guaranteed robustness and justifiable implementation effort. Furthermore, it is impossible to quantify disturbances, failures, and upper bounds of uncertainties such that they can be realistically incorporated into the theoretical performance and robustness bounds. Especially system knowledge about a helicopter is insufficient to rely entirely on theoretical considerations. In conclusion, stability proofs are important to show the operating principle of the algorithm, but are only valid for a helicopter on the paper, i.e. usually a set of differential equations, completed by uncertainties.

Traditional procedures can be applied, every level of design is to be verified and compared to the neighboring level. Does the code written reflect the specifications of the software, does the software designed reflect the specifications of the algorithm, is the algorithm suitable for the control goal, and so on. For further reading about software verification see the V-model in [35] or [36]. An introduction to safety critical systems with $\mathcal{L}_1$-control is given in [37].

6.2 Core Reasoning

The certification strategy proposed here can be outlined as follows:

1. A controller is proposed which evolved out of theoretical insight and during long development phases with lessons learned.

2. Formal (theoretical) verification is conducted, that shows the sophisticated and deterministic approach.

3. Practical verification confirms the predicted performance and robustness. This is mostly testing over the envelope and Monte Carlo runs.

The key of verification lies in the complement of theoretical and practical components. Sophistication means that the controller does not work by coincidence but follows a deliberate procedure. It further extends to what is shown in the cause and effect analysis in the appendix. The choice of design parameters thus has known effects.

For nonlinear and strongly time varying systems, the time delay margin is currently the acknowledged robustness metric. In [13], a conservative upper bound for the time delay margin is computed analytically, however only for a specific linear system. Because of this and by requiring too much system knowledge it is not used in this thesis. Hence, the time delay margin is determined experimentally.

The paper [36] suggests that in general adaptive software is more difficult to bring into accord with DO-178(B) in terms of definition of software (performance) requirements and verification – among few others. With the discrete and deterministic functionality of the controllers shown in this thesis, the $\mathcal{L}_1$-controllers should be able to be verified similar to PI controllers. The same holds for software performance requirements. An implementation on the target flight control computer can be compared to the implementation on the simulator.

The proposed $\mathcal{L}_1$-controller is designed as deterministic, algorithmic computing unit, in contrast to evolving software that is not explicitly coded e.g. artificial neural networks, cf. [34]. Furthermore, the idea of linear systems to extrapolate local performance to the performance near this local point can be applied to the $\mathcal{L}_1$-control architectures shown. The $\mathcal{L}_1$-controllers shown are LTI, or LTV if scheduled. The proofs in Appendices E and F show that performance is "convex" with the time step length and the magnitude of uncertainties/disturbances. In other words, the smaller the time step and the smaller and slower changing the uncertainties/disturbances, the better the performance. If performance can be shown at the limits of the envelope, similar or better performance can be expected within the envelope. The same is true for uncertainties/disturbances in general. Thus, it should be possible to certify the controller analog to current PI-based controllers and DO-178 can be applied as is. However, more emphasis is put on formal mathematics due to a (for most people) less intuitive concept. The philosophy of the mutual endorsement of formal methods and practical testing is therewith not disputed.

To the author's knowledge neither gust charts nor gain and phase margin are incorporated in the current helicopter certification process – supporting the ideas from above. Only border points of the envelope are proven to show satisfying performance and handling qualities. Current requirements as the European specifications CS-27 or CS-29 demand sufficient damping however.

Other arguments may based on the deterministic workflow in combination with static structures.

> The piece-wise constant adaptive law is not an online identification, but a deterministic strategy to cancel the prediction error – the adaptive law is a constant gain (or slowly time varying if scheduled).

A certification approach for adaptive systems with the focus on online learning can be much more tedious. See for example [38] or [39].

6.3 Additional Supportive Criterions

All general requirements of Chapter 1.2 have to be fulfilled for certifiable algorithms. Taken together:

- The filtering structure prohibits high frequencies of estimated uncertainties to be propagated to the input channel.

- The controller *structure* is not adaptive.

- Measurement noise is attenuated as shown in Appendix J.

- Chapter 3 shows that the input channel structure protects against drift between predictor states and plant states due to input saturation.

- An $\mathcal{L}_1$-controller with the piece-wise constant adaptive law uses the same elements as the baseline controller. These are discretely implemented filters of low order, integrators, limiters (saturation), addition, multiplication and in few cases (some scheduling methods) scalar division. The number of these elements is close to the elements used in the baseline controller. It may even be possible to reduce software complexity with fewer lines of code to be implemented.

- The biggest number that is hard-coded is the adaptive gain, see Figure 6.1. The necessity of larger numbers depends on the unit of $\tilde{x}(t)$ which gets multiplied by this gain, and $\tilde{x}(t) < (\ll)1$, if $x(t)$ $[rad/s]$.

- The controller is deterministic. The "learning" is an inversion of the error dynamics. This inversion drives some terms of the error dynamics to zero after one T, thus explicitly including T and repeating it at every iT, $i = 1, 2, \dots$. For the real application it must be shown that the time step length is a) small enough for sufficient performance and b) long enough for all computations to be finished with some margin. The fact that the task is repeated over and over again within the same time step length as in a clockwork is one of the most important features of the piece-wise constant adaptive law. Loops with a dynamic number of executions are avoided. In addition, the controller is robust against some amount of deviation of the actual CPU step length from the hard-coded value of T, see Appendix I for more.

- Stability and performance proofs with formal mathematics exists, see Appendices E and F.

- As shown in Appendices E and F, performance is locally guaranteed during the transient.

- A meaningful robustness metric is the time delay margin. If this margin approaches zero, oscillations occur.

- Time delays in the input channel are explicitly taken into account.

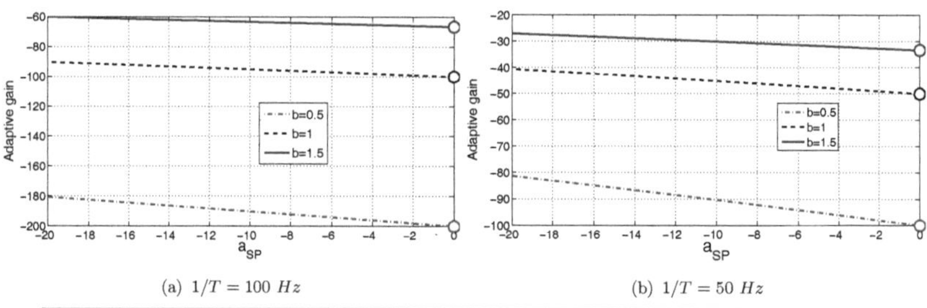

(a) $1/T = 100\ Hz$ (b) $1/T = 50\ Hz$

FIGURE 6.1: Effect of a_{SP} on the size of the adaptive law (the law is a gain)

Sufficient preconditions are fulfilled for a seminal level of acceptance by positively answering all systemic requirements. It remains to be shown that the algorithm is properly designed and tuned, and suitable to the control problem in the first place.

Note that the decoupling of adaptation from the control loop in terms of frequency scales is the important driver for a possible certification. This decoupling allows for separate tuning of adaptation and of control. Most notably, it is up to the engineer to determine the frequency content that is allowed to enter the input channel, defining "aggressiveness" of control, with actuators properties in mind – and the wear of those.

The formal stability proof delivers performance bounds. As mentioned above, these are hard up to impossible to verify with the real ones.
Robustness however is to be determined completely experimentally (mainly in simulations) by adding (or removing as far as possible) time delay solely to the input channel of the plant.

A selection of verification examples can be found in Appendix N.

The final point to demonstrate is the actual performance, mainly referring to sufficient control authority and damping.

Chapter 7

Simulation Results

This chapter provides some simulation results of the primary architecture. A selection of certain benchmarks regarding performance and robustness are shown. For the simulation setup, see Appendix O.

If not stated otherwise, the following sampling times are used. These are the sampling times whose reduction does not provide significant benefits.

State- / Output-feedback in pitch, roll, yaw: $T = 1/(50\ Hz)$

State feedback, vertical speed controller: $T = 1/(100\ Hz)$

For pitch, roll, yaw, the following parameters apply: 60 kts forward speed, MTOM, CG aft limit, standard atmospheric conditions at mean sea level; simulations are conducted without modeled sensor noise, but with notch filters similar to the one in Figure 2.9 (included in $C(s)$ of the $\mathcal{L}_1$-controller or at the baseline controller output) for avoiding excitation of structural resonances.

Controller Parameters in State Feedback (Pitch, Roll, Yaw):
The filters $C_k(s) = \frac{1}{(1/(B_k)s+1)} \frac{1}{(1/(1.2B_k)s+1)}$ (k defining the axis) are implemented with $B_k = 20\ rad/s$ for pitch and yaw, and $B_k = 15\ rad/s$ for roll. Values of $a_{SP} = -0.1$ are chosen. The desired parameters of augmentation are given in Figure O.3. The parameters of the $\mathcal{L}_1$-standalone are : $a_{pitch} = -4.8$, $a_{roll} = -10.6$; $b_{pitch} = 0.6$, $b_{roll} = 1.7$; yaw is scheduled as shown in Figure O.3 e) and f), but a_{yaw} and b_{yaw} of Figure O.3 e) and f) are multiplied by a factor 1.2.

Controller Parameters in Output Feedback (Pitch, Roll, Yaw):
The filters $C_k(s) = \frac{1}{(1/(B_k)s+1)} \frac{1}{(1/(1.2B_k)s+1)}$ with $B_k = 20\ rad/s$ are used for every axis. The

desired transfer functions are $k\frac{(s+20)}{(s+10)(s+15)}$, where k is 1.25, 0.19, 0.27 for pitch, roll, yaw.

Controller Parameters in State Feedback (Vertical Speed):
The filter $C(s) = \frac{1}{(1/(B)s+1)}\frac{1}{(1/(1.2B)s+1)}$ is implemented with $B = 25 \ rad/s$. The desired dynamics are defined by $a = -1$, $b = -0.45$; A value of $a_{SP} = -0.01$ is chosen.

- Output feedback is shown with the recursive adaptive law only.
For standalone $\mathcal{L}_1$-controllers in pitch, roll, and yaw, the outer attitude loops are active.
- The baseline controller alone is evaluated by closed-loop system identification and feedback analysis with *computed* sensitivity functions and linear margins.
- For the augmentation and the standalone $\mathcal{L}_1$-controllers, the sensitivity functions are obtained *experimentally* by a frequency sweep applied to the actuator command, acting as input disturbance $d_{in}(t)$. The identification is conducted between the frequency sweep as input and the actuator command vector (after adding the sweep to the actuator command vector) as output.
- Gain and phase margin are not computed for the $\mathcal{L}_1$-controllers as this is possible only in exceptional cases or by the linear limiting behavior of the controller, cf. [15].
- The coherence ("measurement for linearity" in linear system identification) is not presented as it has been shown to be sufficiently high in all simulations, namely throughout above 0.6, but usually between 0.8 and 1.
- The definition of the desired dynamics is verified or falsified in Appendix N for augmentation, where an overshoot is introduced into the desired dynamics. **Thus, the results for augmentation are an example of poor performance and are to be compared to the standalone $\mathcal{L}_1$-controller.**
- The terminology used describes the Euler angles as Θ, Φ, Ψ for the attitudes and the body-fixed rates as q, p, r, both for pitch, roll, yaw.
- "Actuator command" describes the command *for* the actuators.
- Time delay margins shown are only locally valid (as is the gain and phase margin in PI controllers), as they strongly depend on the flight condition and other parameters. By manually examining the flight envelope, a lower bound (thus a conservative bound) of the time delay margin is searched for, i.e. the additional time delay which shows oscillations of constant amplitude in at least one flight condition.

All simulation results are *one* solution. Regardless of the controller architecture, different results can be achieved. The trade-off between performance and robustness allows for variations.

7.1 Frequency Domain: Baseline Controller

Similar results are shown in [3] and [1].

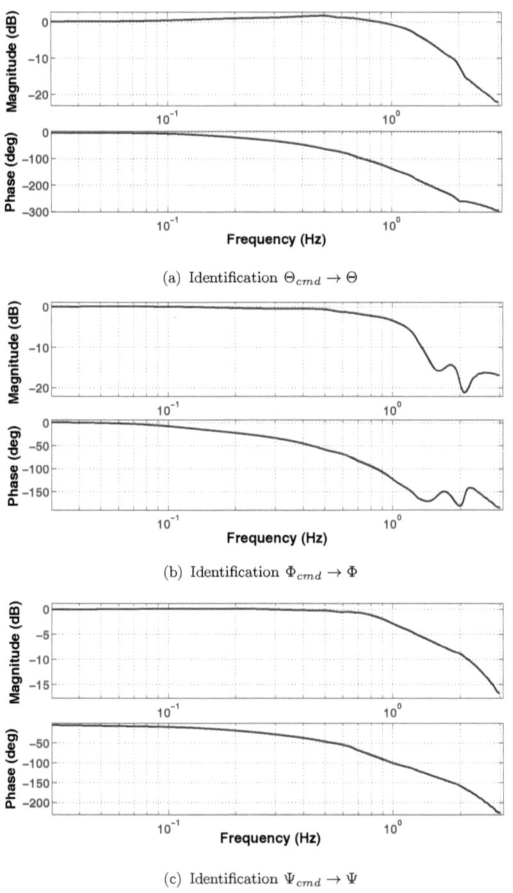

(a) Identification $\Theta_{cmd} \to \Theta$

(b) Identification $\Phi_{cmd} \to \Phi$

(c) Identification $\Psi_{cmd} \to \Psi$

FIGURE 7.1: Attitude tracking of the baseline controller

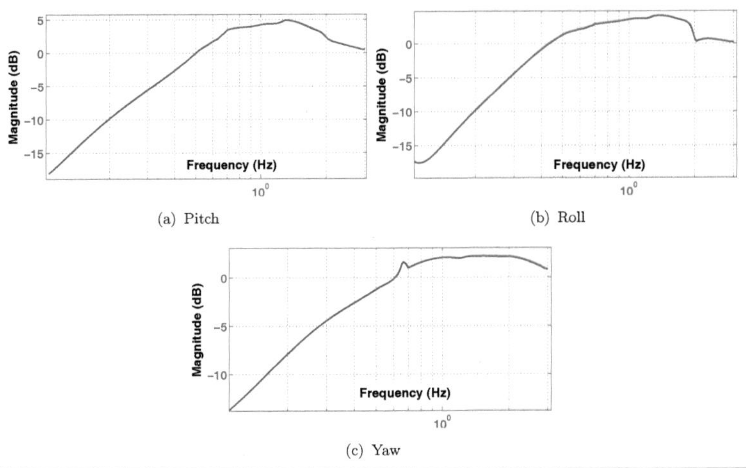

(a) Pitch

(b) Roll

(c) Yaw

FIGURE 7.2: Analytically computed sensitivity functions of the baseline controller

	Pitch	Roll	Yaw
Gain(dB)/Phase(deg) margin estimates:	8.61/48.3	8.96/51.2	14.8/69.2

7.2 Frequency Domain: Augmentation with State Feedback

Similar results are shown in [1].

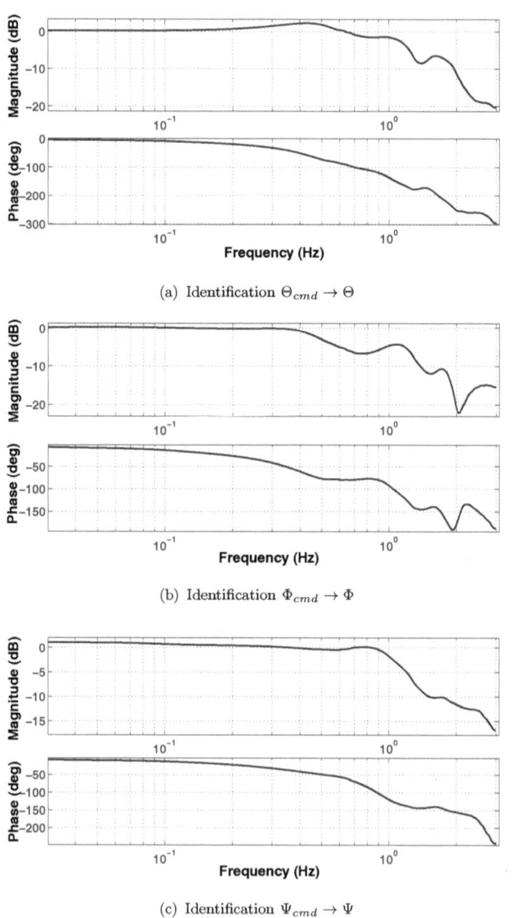

(a) Identification $\Theta_{cmd} \to \Theta$

(b) Identification $\Phi_{cmd} \to \Phi$

(c) Identification $\Psi_{cmd} \to \Psi$

FIGURE 7.3: Attitude tracking of the augmented baseline controller

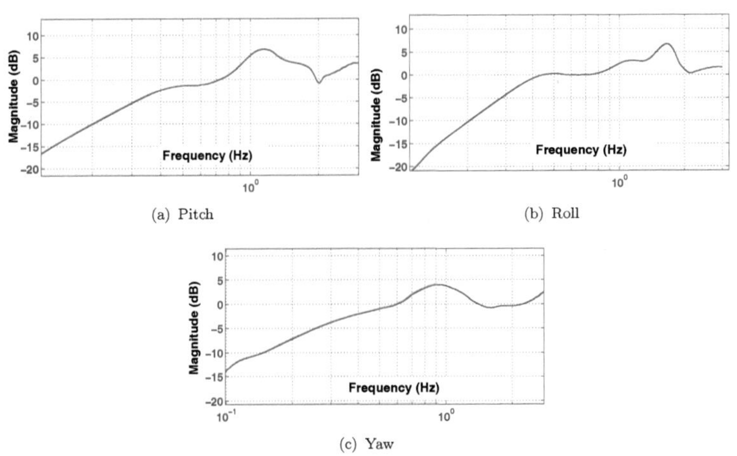

FIGURE 7.4: Sensitivity functions of the augmented baseline controller

	Pitch	Roll	Yaw
Time delay (ms) margin estimates:	40	90	50

7.3 Frequency Domain: Standalone $\mathcal{L}_1$-Control in State Feedback

7.3.1 Raw Adaptive Law

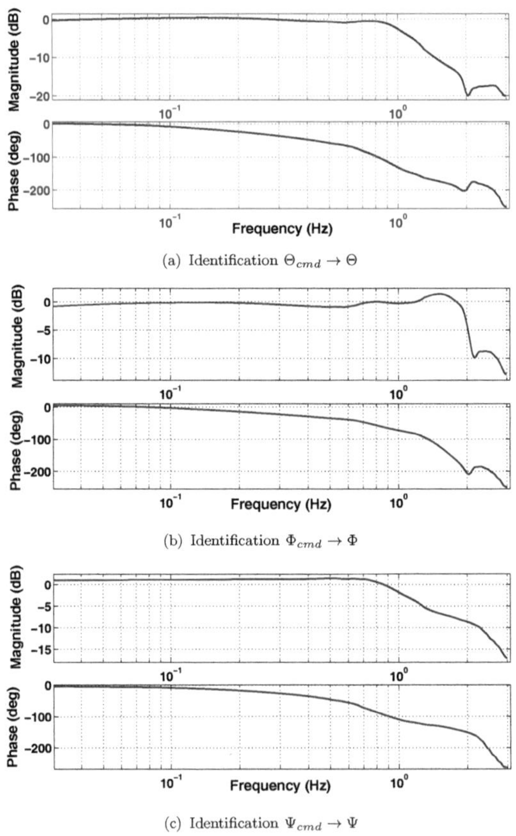

(a) Identification $\Theta_{cmd} \to \Theta$

(b) Identification $\Phi_{cmd} \to \Phi$

(c) Identification $\Psi_{cmd} \to \Psi$

FIGURE 7.5: Attitude tracking of the standalone state feedback controller – raw adaptive law

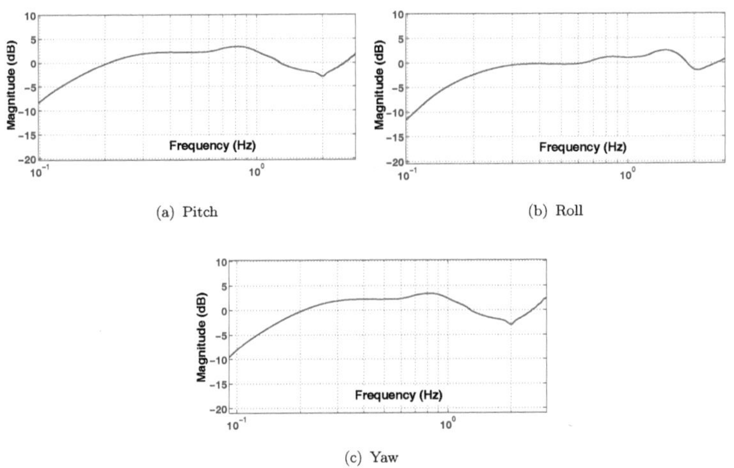

(a) Pitch

(b) Roll

(c) Yaw

FIGURE 7.6: Identification $d_{in}(t) \to u(t)$ in state feedback – raw adaptive law

	Pitch	Roll	Yaw
Time delay (ms) margin estimates:	150	200	50

7.3.2 Recursive Adaptive Law

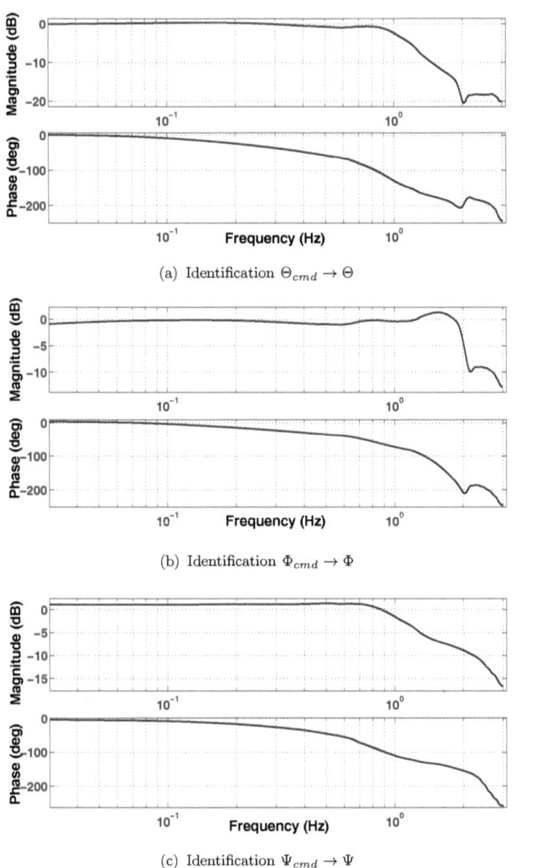

(a) Identification $\Theta_{cmd} \rightarrow \Theta$

(b) Identification $\Phi_{cmd} \rightarrow \Phi$

(c) Identification $\Psi_{cmd} \rightarrow \Psi$

FIGURE 7.7: Attitude tracking of the standalone state feedback controller – recursive adaptive law

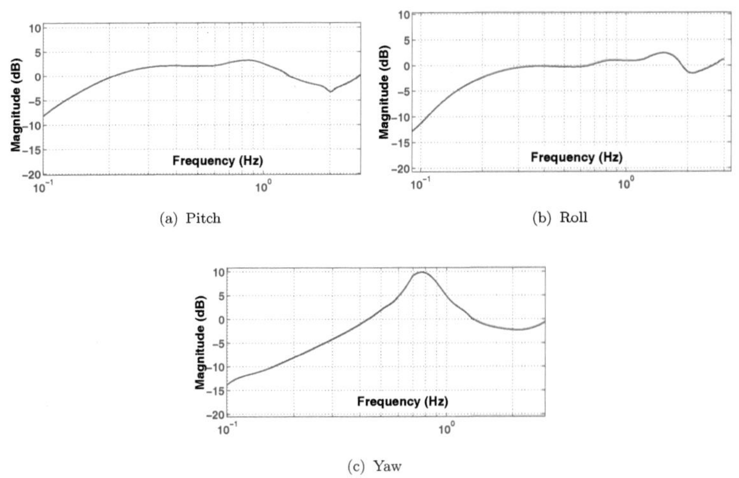

FIGURE 7.8: Identification $d_{in}(t) \to u(t)$ in state feedback – recursive adaptive law

	Pitch	Roll	Yaw
Time delay (ms) margin estimates:	140	150	50

7.4 Frequency Domain: Standalone $\mathcal{L}_1$-Control in Output Feedback

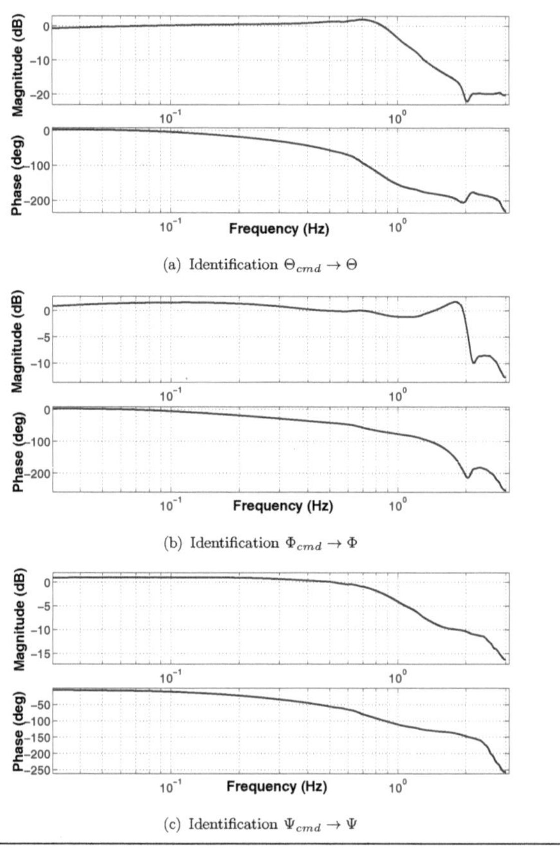

(a) Identification $\Theta_{cmd} \to \Theta$

(b) Identification $\Phi_{cmd} \to \Phi$

(c) Identification $\Psi_{cmd} \to \Psi$

FIGURE 7.9: Attitude tracking of the standalone output feedback controller – recursive adaptive law

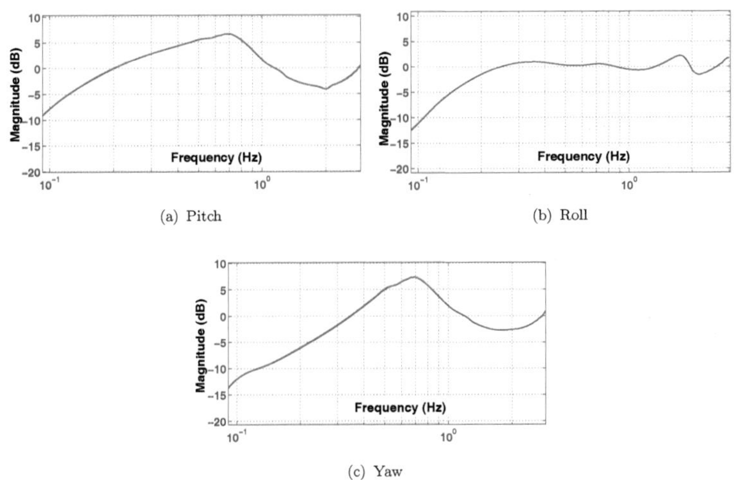

(a) Pitch

(b) Roll

(c) Yaw

FIGURE 7.10: Identification $d_{in}(t) \to u(t)$ in output feedback – recursive adaptive law

	Pitch	Roll	Yaw
Time delay (ms) margin estimates:	100	170	50

7.5 Time Domain: Standalone $\mathcal{L}_1$-Control in State Feedback – Raw Adaptive Law

Flight condition: Hover (slightly out of ground effect).

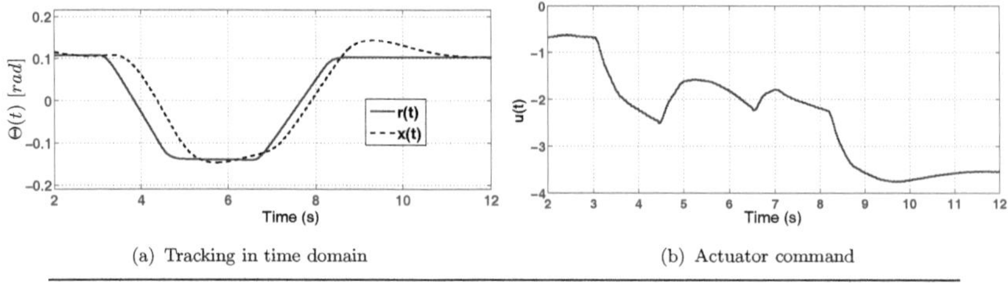

(a) Tracking in time domain   (b) Actuator command

FIGURE 7.11: Attitude tracking in time domain and actuator history – pitch

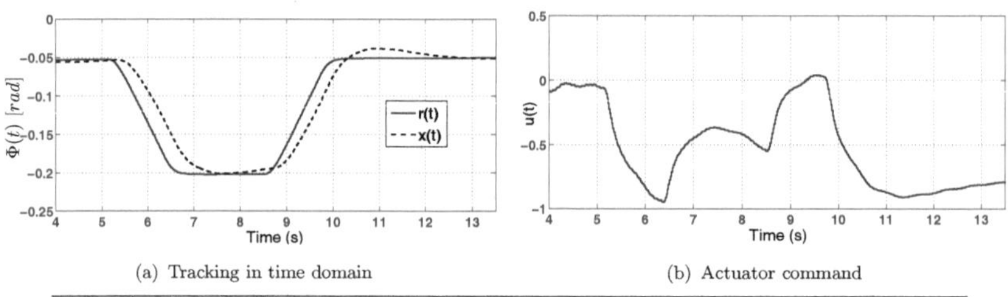

(a) Tracking in time domain   (b) Actuator command

FIGURE 7.12: Attitude tracking in time domain and actuator history – roll

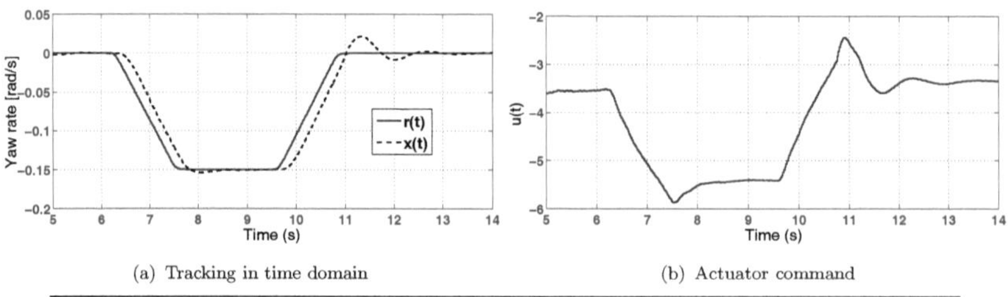

(a) Tracking in time domain   (b) Actuator command

FIGURE 7.13: Rate tracking in time domain and actuator history – yaw

7.6 Vertical Speed Controller

Similar results were first shown in [1] and (with these figures) in [2].

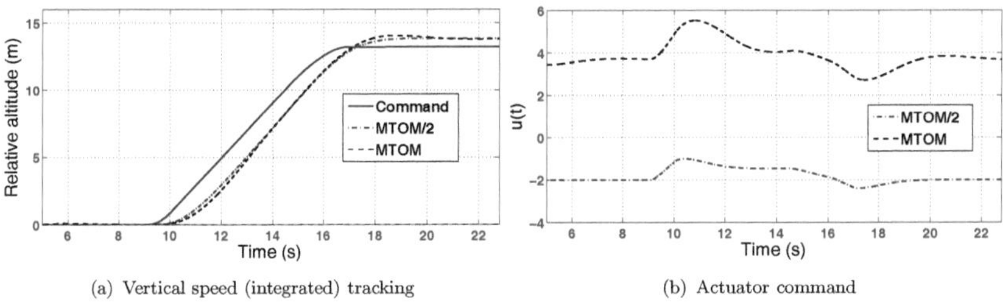

FIGURE 7.14: Vertical speed over time (integrated) for MTOM and MTOM/2 – *without* actuator position saturation

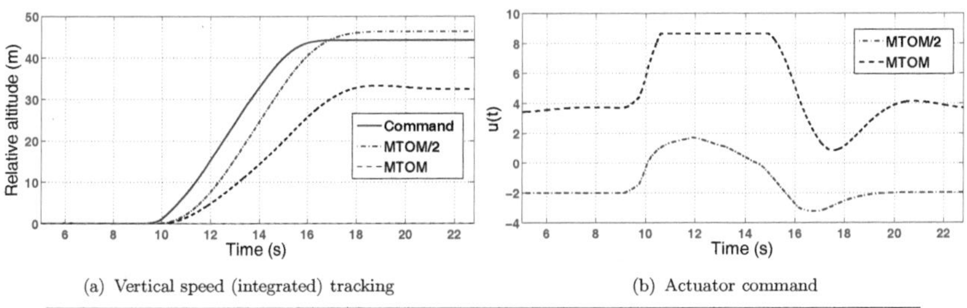

FIGURE 7.15: Vertical speed over time (integrated) for MTOM and MTOM/2 – *with* actuator position saturation

Note that in Figure 7.15 a higher altitude difference in the same time as in Figure 7.14 is commanded. With a higher vertical speed, the actuator saturates with MTOM. The integrated vertical speed (altitude) is not targeted, only the behavior in the vertical speed is sought.

The mass MTOM/2 is slightly below empty weight, which demonstrates some margin as lower helicopter masses prove to be more prone to control signal oscillations than higher ones.

	MTOM	MTOM/2
Time delay (ms) margin estimates:	600	400

Chapter 8

Conclusion and Future Work

8.1 Conclusion

This thesis shows various $\mathcal{L}_1$-controllers for helicopters in pitch, roll, yaw, and for vertical speed in low-speed regimes. For attitude control, combinations of state and output feedback with standalone and augmenting $\mathcal{L}_1$-controllers are shown, apart from the augmentation of the baseline controller in output feedback, which is considered overly complex and to provide only moot benefits.

$\mathcal{L}_1$**-control is a fast adapting model following strategy** with a complementary high-pass – low-pass structure. While the low-pass filtering structure ensures frequency content appropriate for the input channel of the plant, the high-pass structure excludes high frequency uncertainties from the desired behavior – the prediction error remains small. Applying this complementary filtering structure allows for fast adaptation without sacrificing robustness (cf. [13]). Either a digital command or the baseline controller input containing the command can be sent to the predictor.

Saturation protection is guaranteed for inputs shared by the plant and the predictor, and for modified predictor inputs, e.g. with hedging of trim or decoupling signals. Rigorous conditions are shown.

The input channel to the predictor is an important tuning element: Saturation, time delay, actuator dynamics, or a simple model of the rotor response are vital elements.

There are **fundamental differences between the baseline controller and** $\mathcal{L}_1$**-control:** Where loop shaping of the baseline controller may be one of the most appropriate and systematic techniques currently available for helicopters, the $\mathcal{L}_1$-controller can be seen as a model following controller. In other words, the $\mathcal{L}_1$-controller may be stricter, especially regarding the recovery of the control effectiveness. Thus, scheduling can be necessary for the $\mathcal{L}_1$-controller where for the baseline controller it is not.

The gap between frequency domain design of the baseline controller and the design and tuning of the $\mathcal{L}_1$-controller is narrowed by the complementary filter structure of $\mathcal{L}_1$-control. Frequency domain is acknowledged to be often useful for benchmarks, e.g. in ADS-33 ([4]). The $\mathcal{L}_1$-controller requires less system knowledge due to its adaptive tracking strategy.

Several options to reduce the propagation of undesired dynamics and disturbances to the tracking error exist for the $\mathcal{L}_1$-controller:

Using a baseline controller – be it just for providing a trim signal – reduces the deviation of the desired from the undesired dynamics; reducing the sample time reduces the integration time of these undesired dynamics; applying a memorizing term helps to "learn" the uncertainties and to counter their integration within a computational time step; choosing slow error dynamics in scalar state feedback helps effectively; it is shown that *slower* error dynamics are better performing, but are less robust (valid for a scalar system with the piece-wise constant adaptive law).

> Choosing slow a_{SP} in state feedback and making use of the recursive piece-wise constant adaptive law in output feedback turns out to be the key for satisfying performance in this thesis.

Standalone $\mathcal{L}_1$-control has several benefits over augmentation:

Augmentation requires the baseline controller to define desired dynamics together with the predictor dynamics – a demand that is hard to achieve with a higher order baseline controller designed in loop shaping. With these preconditions, augmentation is susceptible to introduce undesired dynamics.

Predictor dynamics show to be much more free to choose for the standalone controller, i.e. also slower than the open-loop dynamics. In augmentation however, the predictor dynamics have to be very close to the open-loop dynamics. Augmentation may hurt the time delay margin. Besides that, not modifying the baseline controller seems hardly possible as anti-wind-up strategies in the baseline controller do not account for the adaptive control signal by default.

A vertical speed controller in state feedback for hover and low-speed regimes is proposed. A special structure for trim subtraction with a dead zone and hysteresis combined with a smooth transition is proposed. A mass estimator is eluded by a robust design of the maximally expected mass changes. Agility in case of flight with low masses is not deteriorated with desired dynamics oriented at empty weight and a robust controller design.

Special emphasis in this thesis is put on the application to commercial systems. This embraces portability between various helicopter types with little retuning effort as well as code complexity. When porting code, adjustments in tuning are potentially required in mostly the low-pass filter bandwidth, the desired dynamics, the modeled time delay. The reason for this is the internal model based architecture, and the "adaptive" strategy of achieving it. Proper saturation and the explicit inclusion of the limited input channel bandwidth support this scheme.

The controller however fits best for fly-by-wire architectures. The major benefit of the proposed architectures may be seen in reduced development effort, as iterative tuning with system identification can be reduced.

Proofs of performance bounds are provided with formal mathematical strategies. A different structure in state feedback, the inclusion of sensor noise in all proofs, and a recursive adaptive law with an integrated initialization procedure in output feedback are the main features of these proofs (shown in the Appendix).

Equivalent forms of the state predictor are shown in the Appendix. Benefits are discussed. The proofs however are not used for design, their only role is to provide insight and the demonstration of a deterministic and sophisticated control strategy.

A certification strategy may be based on the duality of theoretical proofs and verification, insights as well as the final testing. The available amount of system knowledge about a helicopter is insufficient to bypass extensive testing in high-fidelity simulations like Monte Carlo runs and finally real flight tests, regardless of which control paradigm is applied. The key to certification is the deterministic mode of operation with the discrete nature of the piece-wise constant adaptive law.

8.2 Future Work

This section provides an outlook on potentially envisaged continuations.

- It may be possible to further tweak the filtering structure, e.g. with the introduction of zeros.

- If required, a multirate $\mathcal{L}_1$-controller can be implemented. That means that the loop "predictor – adaptive law" uses different time steps than e.g. the sensor signal is provided with. In the piece-wise constant adaptive law, the faster CPU frequency keeps the integration time of the cumulated uncertainty lower despite the fact that within several CPU cycles new sensor information is not provided.

- It may become necessary to schedule the modeled time delays.

- State feedback: Instead of a linear predictor, one in the form of $\dot{x}(t) = f(x(t), t) + g(x(t), t)u(t) + a_{SP}\tilde{x}(t)$ may be introduced. If in the linear predictor $ax(t)$ is replaced with a nonlinear $f(x, t)$, the adaptive law can be used as is, since $f(x(t), t)$ does not appear in the error dynamics, which are still determined by a_{SP}, b, and T. The main challenge for nonlinear desired dynamics lies in the stability proof as the $\mathcal{L}_1$-norm for a general $f(x(t), t)$ potentially cannot be computed straightforwardly. If b is replaced with $g(x(t), t)$, a time varying $b(t)$ in the adaptive law is to be designed as a time varying linearization of this function $g(x(t), t)$. More insights to nonlinear desired dynamics can be found in [40].

- State feedback: If signal quality allows, angular accelerations can be included in the predictor dynamics. If rates are used additionally, the adaption may not be unique anymore as the rates are an uncertainty free integration of the angular accelerations.

- Output feedback: The output feedback controller shown in this thesis can be modified according to [22]. In this paper, state dependent as opposed to output dependent nonlinearities are included. The implementation however is very similar. The proof is valid for a broader class of systems and constraints for the desired transfer function may relax.

- The function of the recursive adaptive law can be replaced with a memorizing term in the predictor. This architecture is shown in [41].

Appendix A

Definitions

A.1　General Definitions

Definition A.1.1.

The term **actuator** refers to active control units – although the understanding of the rotor being an actuator is justified, the terminology shall refer to hydraulic, electrohydraulic, electric etc. actuators.

Definition A.1.2.

The **bandwidth** of a system is the frequency where a magnitude loss of $3dB$ of the DC-gain occurs.

Remark A.1.1.

For helicopters, different definitions of "bandwidth" can apply. See for instance [4] (ADS-33).

Definition A.1.3.

Target signals are denoted as **command** (instead of reference).

Definition A.1.4.

The **reference system** is a virtual closed-loop plant. A hypothetic $\mathcal{L}_1$-controller with an ideal adaptive law closes the loop, i.e. the uncertainty is known. This system is not real but used for analysis.

Definition A.1.5.

The **desired system** states user defined dynamics, targeted as ultimate control objective.

Definition A.1.6.

A **discrete system** changes only at fixed and equidistant points in time.

Definition A.1.7.

A **quantized system** has limited numeric precision.

Definition A.1.8.

Low-frequency pertains to frequencies within the bandwidth of the system input channel, **high-frequency** to frequencies beyond.

Definition A.1.9.

A **Pareto front** is a set of design parameters which are "equally" optimal regarding the cost function. This term is used in muli-criteria optimization if several sub- cost functions are added to a total cost function to be minimized. It is not possible to reduce one of the cost functions without increasing another one when moving on the Pareto front.

Definition A.1.10.

The **on-axis** describes in system identification the same axis where the input is applied. The **off-axes** are all axes not being the on-axis but potentially coupled to the input of the on-axis.

Definition A.1.11.

The **safe flight envelope (SFE)** defines the technical bound of flight conditions where the vehicle reaches its physical limits, e.g. structural loads.

Definition A.1.12.

The **operational flight envelope (OFE)** is a user-defined bound for flight conditions the pilot does not exceed during normal operations, always within the SFE but with sufficient margin to the bounds of the SFE.

Definition A.1.13.

A Lipschitz constant is a measurement for the change rate of a function, in scalar systems an upper bound for its slope. A function $f : \mathbb{R} \to \mathbb{R}$ has the Lipschitz constant $L > 0$ such that $\frac{|f(x_2)-f(x_1)|}{|x_2-x_1|} \leq L$. Analog for $f : \mathbb{R}^n \to \mathbb{R}^n$: $\|f(x_2) - f(x_1)\| \leq \|x_2 - x_1\|$.

Definition A.1.14.

A **matched uncertainty** is an uncertainty or disturbance that can be simulated in the input channel. An uncertainty is either matched or unmatched, but not both.

Remark A.1.2.

Definition A.1.14 is consistent with the fact that for baseline controller tuning the sensitivity function (gained from the closed-loop transfer function analytically) is verified by applying a frequency sweep as disturbance at the actuator input. This is true for linear SISO systems.

"Unmatched uncertainty" means, an input matrix B is such that for this uncertainty B lacks row entries for a respective input. For instance, if in a distributed system drone #1 controls the position relative to drone #2, then movements of drone #2 are an unmatched uncertainty for drone #1. Another example are uncertainties in the angle of attack of a fixed wing plane while controlling the short period via pitch rate.

Definition A.1.15.

The **time delay margin** is the amount of time delay that is added in the input channel of the plant for the system to become marginally stable, i.e. the output as response to an input or a disturbance shows oscillations with constant amplitude over time. In other words, the time delay margin quantifies the amount of time delay which, when added to the plant input, causes zero phase margin.

Remark A.1.3.

Subtracting a certain amount of time delay from the input channel can be possible to some extent in simulations. In combination with adding time delay, it helps to a more accurate understanding of robustness.

Definition A.1.16.

An LTI transfer function $H(s)$ is called **unstable**, if one or more poles lie in the right half plane, it is called **marginally stable** if none of the poles lies in the strictly positive right half plane but some on the imaginary axis, and it is called **stable** if all real values are strictly negative.

Definition A.1.17.

An LTI transfer function $H(s)$ is called **proper**, if $H(\infty)$ is finite, i.e. $n_{Poles} \geq n_{Zeros}$, and strictly proper if $H(\infty) = 0$, i.e. $n_{Poles} > n_{Zeros}$.

Definition A.1.18.

A transfer function is minimum phase if its numerator is Hurwitz.

Definition A.1.19.

A transfer function $H(s)$ is called **positive real** if $RE\,[H(s)] \geq 0 \; \forall \; RE\,[s] \geq 0$, where $RE[\cdot]$ is the real part.

A transfer function $H(s)$ is called **strictly positive real** (SPR) if $H(s - \epsilon)$ is positive real for $\epsilon > 0$.

Remark A.1.4.

Definition A.1.19 implies necessary conditions on a transfer function $H(s)$ to be SPR: $H(s)$ is stable, $H(s)$ is strictly minimum phase, the relative degree of $H(s)$ is 0 or 1, the phase lag is always less than 90 *deg*.

Definition A.1.20.

A matrix A is **Hurwitz** if all its eigenvalues have negative real part.

Definition A.1.21.

A matrix A **is positive definite** $(A > 0)$, if $x^T A x > 0 \; \forall x \neq 0 \; AND \; x^T A x = 0$ for $x = 0, \quad x \in \mathbb{R}^n$.

Definition A.1.22.

A function $f : \mathcal{D} \to \mathbb{R}, \; f(0) = 0$ is **positive definite**, if $f(x) > 0, \quad x \in \mathcal{D} \setminus \{0\}$.

Remark A.1.5.

There are many definitions of **stability** in dynamic systems, e.g. exponential decay in linear systems, $\mathcal{L}_p$-stability, or Lyapunov-stability for more general nonlinear systems. The reader is advised to consult respective textbooks.

Definition A.1.23.

A set Ω is **positively invariant**, if from $x(0) \in \Omega$ follows that $x(t) \in \Omega \; \forall t \geq t_0$, while $x \in \mathbb{R}^n$. Trajectories can in case of an invariant set neither enter nor leave it, in case of a positively invariant set enter but not leave.

Definition A.1.24.

The 2-norm and ∞-norm of a vector $f \in \mathbb{R}^n$ are defined as:

$$\|f\|_2 := \left(f^T f\right)^{1/2}$$
$$\|f\|_\infty := \max_{1 \leq i \leq n} \left(|f_i|\right)$$

Definition A.1.25.

The $\mathcal{L}_1$-norm, $\mathcal{L}_2$-norm, and $\mathcal{L}_\infty$-norm of a function $f(t)$ are defined as:

$$\|f\|_{\mathcal{L}_1} := \int_{\mathcal{I}}^{\infty} \|f(\tau)\| \, d\tau < \infty$$
$$\|f\|_{\mathcal{L}_2} := \int_{\mathcal{I}}^{\infty} \left(\|f(\tau)\|^2 \, d\tau\right)^{1/2} < \infty$$
$$\|f\|_{\mathcal{L}_\infty} := \max_{1 \leq i \leq n} \left(\sup_{\tau \geq 0} |f_i(\tau)|\right) < \infty$$

where $\|\cdot\|$ can be an arbitrary norm and $\mathcal{I}$ is the absolute zero in time (to avoid confusion in the formal proofs where some integrals reach from a relative "0" to T).

The $\mathcal{L}_1$-norm of transfer functions is defined as $\mathcal{L}_1$-norm of the impulse response, i.e. it is defined in time domain. With $h(t)$ denoting the impulse response in time domain of $H(s)$, $\|h\|_{\mathcal{L}_1}$ is the simplified notation for $\|H(s)\|_{\mathcal{L}_1}$.

Remark A.1.6.

Using the $\mathcal{L}_1$-norm is advantageous to describe the transient. For instance, a slender and high peak appears much better in the $\mathcal{L}_1$-norm than in the $\mathcal{L}_2$-norm of the signal. The $\mathcal{L}_2$-norm however may be well suited to describe signal power.

Definition A.1.26.

The extended $\mathcal{L}_p$-space is defined as $\mathcal{L}_{p,e} := \{u \mid u_\lambda(t) \in \mathcal{L}_p, \ \forall \lambda \in [0, \infty)\}$ with $u_\lambda(t)$ being the truncated function $u(t)$, such that $u_\lambda(t) = u(t)$ for $0 \leq t \leq \lambda$ and $u_\lambda(t) = 0$ for $t > \lambda$.

Remark A.1.7.

The extended space is less restrictive. A broader class of functions belongs to the $\mathcal{L}_{p,e}$-space. The principle of equivalence (=if one $\mathcal{L}$-norm exits, every other does, too) still does not hold for functions.

A.2 Coordinate Frames

In this thesis, three Cartesian coordinate systems are used. The terms "pitch", "roll", "yaw" are turns around the body-fixed y-, x-, and z-axis. The attitude is described by Euler angles Ψ, Θ, Φ. They appear in the order yaw, pitch, roll between an NED and the B-frame.

Body-Fixed Frame. The origin of the body-fixed frame ("B-frame") resides at a fixed distance from some position on the fuselage. The exact location of the origin is not of importance as mainly body-fixed turn rates and attitudes are of concern.

The x-axis points in the direction from tail to nose of the helicopter. The y-axis, perpendicular to the x-axis, points to the right hand side (in pilots' line of gaze) and the z-axis downwards.

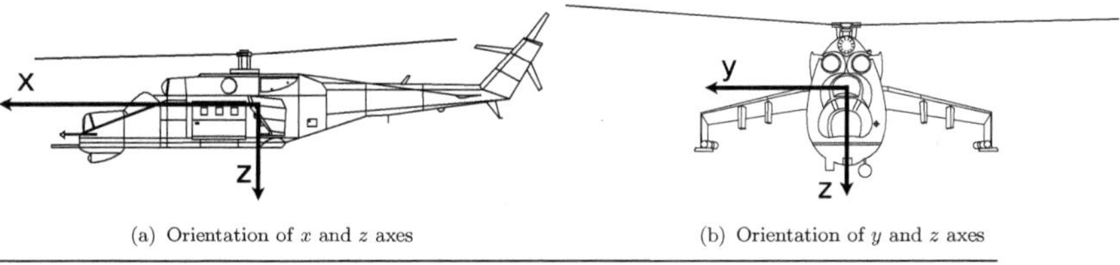

(a) Orientation of x and z axes　　　　(b) Orientation of y and z axes

FIGURE A.1: Axes orientation in the body-fixed coordinate frame

North-East-Down Frame. The north-east-down frame ("NED-frame") shares its origin with the one of the B-frame. Its x-axis points north, its y-axis east and the z-axis is the normal vector of a tangential plane to the surface of an idealized (i.e. perfectly ellipsoid) world, pointing downwards.

Inertial Frame. By neglecting any velocity or acceleration of the earth, any NED-frame is an inertial frame.

Rotor Frame. This coordinate frame ("R-frame") is the body-fixed frame but spinning with the rotor.

Appendix B

Alternatives for Decoupling of Cross-Couplings in Augmentation

This chapter provides an overview of the decoupling of cross-couplings in the adaptive augmentation of the baseline controller by an $\mathcal{L}_1$-controller. A similar list is also presented in [1] by the author. Various options with their potential advantages "+" and shortcomings "−" are listed in the following table.

Approach 1: Model couplings as non-diagonal state space entries of a MIMO state predictor and cancel those by respective terms in the baseline controller, which are designed for these models. Anything not covered by the model is left to feedback from the controllers. As the gyroscopic effects are greater than the ones of inertia, most of these entries are assumed to originate from gyroscopic effects. The rotor however is not a perfect gyroscope and thus cannot be modeled as one. Moreover, when linearizing nonlinear system equations, the cross-coupling of inertia effects excluding the gyroscopic effects vanish. In system identification however these effects are visible.

+ Fully decoupled desired dynamics.
− Might discard information about the system with insufficient models.
− High design effort to decouple a fully coupled state space system.
− Requires redesign of baseline decoupling functions.
− Hedging can be diluted if the input is defined "after the actuators". Cancellation of coupling signals with inclusion of actuator dynamics is required.

Approach 2: Leave the baseline controller unchanged and let these decoupling functions decouple the predictor, whose couplings are derived from baseline decoupling functions itself. The coupling functions are added as additional terms in the predictor.

+   Not modifying a certified baseline controller is desired for a subsequent augmentation.

+   Fully decoupled desired dynamics.

+   Incorporates all available information about coupling.

−   Uncertain feasibility, high implementation effort. Decoupling functions designed in frequency domain are copied and implemented inverted – if proper after inversion in the first place.

−   Hedging can be diluted if the input is defined "after the actuators". Cancellation of coupling signals with inclusion of actuator dynamics is required.

Approach 3: Similar to Approach 2, but with altered coupling and decoupling term. The couplings are modeled by physical insight as additional term, a baseline controller compensates for these couplings with its inverse and negative sign, respectively.

+   Fully decoupled desired dynamics.

−   Low chances of proper modeling – may lack performance.

−   Hedging can be diluted if the input is defined "after the actuators". Cancellation of coupling signals with inclusion of actuator dynamics is required.

Approach 4: Discard all knowledge about cross-couplings and let baseline and/or adaptive controller solve it by error feedback.

+   Lowest computational effort.

−   May lack performance in terms of sufficient decoupling.

Approach 5: Leave the baseline controller unchanged with its decoupling transfer functions and hedge those from the predictor.

+   Not modifying a certified baseline controller is desired for a subsequent augmentation.

+   Inclusion of all available knowledge to decouple the plant.

+   Fully decoupled desired dynamics.

−   Requires implementation of a hedging structure.

−   Hedging can be diluted if the input is defined "after the actuators". Cancellation of coupling signals with inclusion of actuator dynamics is required.

Appendix C

$\mathcal{L}_\infty$-Stability

C.1   Introduction and Conditions

Let $H(s)$ be a strictly proper and stable transfer function with m inputs and n outputs and $h(t) \in \mathbb{R}^{n \times m}$ its impulse response in time domain. Hence, the $\mathcal{L}_1$-norm of the impulse response exists, $\|h\|_{\mathcal{L}_1} < \infty$. Further, $u(t) \in \mathcal{L}_{\infty,e}$ is a bounded input vector to the transfer function $H(s)$.

Theorem C.1.1. The following holds:

$$\|y_i\|_{\mathcal{L}_\infty} \le \|h\|_{\mathcal{L}_1} \|u_i\|_{\mathcal{L}_\infty} \text{ and } y(t) \in \mathcal{L}_{\infty,e}$$

The proof can be found in section C.2.

C.2   Proof of Theorem C.1.1

This proof in its original version can be found in [13].
Let the $\mathcal{L}_1$-norm of $h(t)$ be:

$$\|h\|_{\mathcal{L}_1} := \max_{i=1\ldots n} \left(\sum_{j=1}^{m} \|h_{ij}\|_{\mathcal{L}_1} \right) \tag{C.1}$$

The convolution of the input $u(t)$ and the function $h(t)$ to the output $y(t)$ is defined as:

$$y(t) = h * u = \int_0^t h(t-\tau)u(\tau)d\tau \tag{C.2}$$

For the ith-element of $y(t)$ and the jth-element of $u(t)$:

$$y_i(t) = \int_{t_0}^t \left(\sum_{j=1}^m h_{ij}(t-\tau)u_j(\tau) \right) d\tau \tag{C.3}$$

Then:

$$
\begin{aligned}
|y_i(t)| &= \left| \int_{t_0}^t \left(\sum_{j=1}^m h_{ij}(t-\tau)u_j(\tau) \right) d\tau \right| \\
&\leq \int_{t_0}^t \left(\sum_{j=1}^m |h_{ij}(t-\tau)u_j(\tau)| \right) d\tau \\
&\leq \int_{t_0}^t \left(\sum_{j=1}^m |h_{ij}(t-\tau)|\,|u_j(\tau)| \right) d\tau \\
&\leq \int_{t_0}^t \sum_{j=1}^m |h_{ij}(t-\tau)| \max_{j=1...m} \left(\sup_{t_0\leq\tau\leq t} |u_j(\tau)| \right) d\tau \\
&= \sum_{j=1}^m \int_{t_0}^t |h_{ij}(\tau)| \max_{j=1...m} \left(\sup_{t_0\leq\tau\leq t} |u_j(\tau)| \right) d\tau \\
&\leq \sum_{j=1}^m \|h_{ij}\|_{\mathcal{L}_1} \|u_l\|_{\mathcal{L}_\infty}
\end{aligned}
\tag{C.4}
$$

Thus:

$$\|y_l\|_{\mathcal{L}_\infty} = \max_{i=1...n} \|y_{i,l}\|_{\mathcal{L}_\infty} \leq \max_{i=1...n} \left(\sum_{j=1}^m \|h_{ij}\|_{\mathcal{L}_1} \right) \|u_l\|_{\mathcal{L}_\infty} = \|h\|_{\mathcal{L}_1} \|u_l\|_{\mathcal{L}_\infty} \tag{C.5}$$

$$\square$$

Appendix D

Equivalent State Feedback Systems

Equivalent forms of a scalar state predictor are shown. An output disturbance with measurement noise is incorporated. The following definitions apply:

$\nu(t)$ — describes external output disturbances and measurement noise;

$u(t)$ — defines the adaptive input $u(t) = u_{\mathcal{L}_1}(t)$

$x(t)$ — defines the real state;

$y(t)$ — defines the measured state affected by disturbances:

$$y(t) = x(t) + \nu(t) = \hat{x}(t) - \tilde{x}(t); \;\; \rightarrow \;\; \tilde{x}(t) = \hat{x}(t) - x(t) - \nu(t);$$

k — is a feedback gain of a proportional baseline controller, representing a baseline controller in general;

a_P — is the assumed system matrix of the open-loop plant;

$a = a_P - bk$ — is the desired system matrix;

a_{SP} — defines eigenvalues for the error dynamics. As a design parameter it can be chosen largely free. Restrictions are shown later;

k_{SP} — is such that the desired eigenvalues of the error dynamics from a_{SP} are obtained, i.e. $a_{SP} = a + k_{SP}$;

Then:

$$\text{Plant:} \quad \dot{x}(t) = ax(t) + b\left(u(t) + \sigma(t)\right) \tag{D.1}$$

$$\text{State predictors:} \quad \dot{\hat{x}}(t) = a\hat{x}(t) + b(u(t) + \hat{\sigma}(t)) + k_{SP}\tilde{x}(t) \tag{D.2}$$

$$\dot{\hat{x}}(t) = ay(t) + b(u(t) + \hat{\sigma}(t)) + a_{SP}\tilde{x}(t) \tag{D.3}$$

$$\dot{\hat{x}}(t) = a_P y(t) + b\left(-ky(t) + u(t) + \hat{\sigma}(t)\right) + a_{SP}\tilde{x}(t) \tag{D.4}$$

$$\dot{\hat{x}}(t) = a_P \hat{x}(t) + b\left(-ky(t) + u(t) + \hat{\sigma}(t)\right) + (a_{SP} - a_P)\tilde{x}(t) \tag{D.5}$$

$$\dot{\hat{x}}(t) = a_P y(t) + b\left(-k\hat{x}(t) + u(t) + \hat{\sigma}(t)\right) + (a_P + k_{SP})\tilde{x}(t) \tag{D.6}$$

$$\text{Error dynamics:} \quad \dot{\tilde{x}}(t) = a_{SP}\tilde{x}(t) + b\tilde{\sigma}(t) + a\nu(t) \tag{D.7}$$

$$\text{Adaptive law:} \quad \hat{\sigma}(iT) = \frac{a_{SP}e^{a_{SP}T}}{b(1 - e^{a_{SP}T})}\tilde{x}(iT) \tag{D.8}$$

$$\text{Control Law:} \quad u(s) = u_{\mathcal{L}_1}(s) = -C(s)\hat{\sigma}(s) \tag{D.9}$$

Analysis implies that all state predictors (D.2)-(D.6) lead to the same error dynamics (D.7). Furthermore, all predictors express the same desired dynamics defined by a.

Note that the adaptive law is determined by the error dynamics a_{SP} independent of the formulation of the predictor parameter a.

Dependent on the purpose and the respective modifications of the predictor, these different forms provide different opportunities. Form (D.3) provides a simple form where input saturations can be easily accounted for, as $u(t)$ is the total and therefore only input signal – as opposed to e.g. $u(t) - bky(t)$. Furthermore, the error dynamics can be designed intuitively by the value of a_{SP}. This structure also provides a convenient way to replace $ax(t)$ with $f(x(t))$, where the error dynamics remain the same.

Appendix E

Performance Bounds in State Feedback

The ideas of the subsequent proof are based on [13], Chapter 3.3. The reader however is not required to be familiar with this book – the following proof tries to be complete. Modifications and additional explanations however apply.

The changes to the proof in [13] comprise:
The proof of this thesis does not consider unmatched uncertainties; a different strategy is conducted in this proof (similar to output feedback shown later); sensor noise is included; a slightly different architecture adding a_{SP} is used; an extension to the adaptive law is shown; the unknown input gain is time varying; the definition of $D(s)$ is dropped in favor of a new $V(s)$ ($C(s)$ is now the implemented filter); a feedforward filter $C_r(s)$ is introduced; several minor modifications or simplifications;

The underlying idea of the proof is this: An unknown plant is controlled. All uncertainties are lumped into one scalar variable per axis. A state predictor, an adaptive law and a filtering structure are the elements of the controller. Boundedness of the prediction error is proved by contradiction due to two mutually dependent semiglobal positively invariant sets. Moving hypothetically to the bound of one of the sets results in the proof of the other invariant set to be true and the exclusion of the bound of the former, which again proofs the original set and contradicts the assumption of the performance bound ever reaching the bound itself.

The reference system serves as intermediate step in calculating the deviation of the desired from the real dynamics.

E.1 Definitions, Assumptions, Descriptions

A discrete controller is applied to a continuous plant. The computation rate, i.e. time step length of the algorithm that is hard-coded in the implementation is denoted as T. It is subject to hardware limitations and chosen as low as task execution time allows for best performance.

The open-loop plant dynamics are modeled as SISO transfer function per axis with unknown dynamics and state dependent nonlinearities. If for future purposes a MIMO formulation were to be needed, the essential terms and results for a MIMO system are shown in framed boxes.

The plant is described by the following set of equations:

$$\dot{x}(t) = ax(t) + b\left(\omega(t)u(t) + f(t, x(t), z(t))\right) = ax(t) + b(\omega_0 u(t) + \sigma(t)) \tag{E.1}$$

$$\dot{x}_z(t) = g(t, x_z(t), x(t)), \quad x_z \in \mathbb{R}^p \tag{E.2}$$

$$z(t) = h(t, x_z(t)) \tag{E.3}$$

This means, $\sigma(t)$ lumps: $f(t, x(t), z(t)) + (\omega(t) - \omega_0)u(t)$.

The meaning of the particular terms is:

$$a, b : \text{A linear plant model or linear desired dynamics}$$

$$f(t, x(t), z(t)) : \text{Lumps all uncertainties, external disturbances affecting } \dot{x}(t), \text{ nonlinearities}$$
$$\text{and effects of internal dynamics}$$

$$(\omega(t) - \omega_0)u(t) : \text{Accounts for uncertainties in the input gain}$$

Despite the measured state $x(t)$ (for MIMO: vector of states $x(t)$), unknown (with assumptions though, see below) internal dynamics are allowed to be present. The output of the internal dynamics affecting the state space model in (E.1) is denoted by $z(t)$, the internal states by $x_z(t)$.

The Laplace transformation of $f(t, x(t), z(t))$ is given by $\eta(s)$.

> For MIMO-systems:
> $$\dot{x}(t) = Ax(t) + B\left(\omega(t)u(t) + f(t, x(t), z(t))\right), \quad f(t, x(t), z(t)) \in \mathbb{R}^n; \qquad y(t) = \mathcal{C}x(t)$$

Define the initial condition for the plant's differential equation: $x_0 := x(t = 0)$

Define the upper bound for x_0 as ρ_0: $\|x_0\|_\infty \leq \rho_0$

Define the initial condition for the Laplace-transformed differential equation: $x_{in}(s) := (s-a)^{-1}x_0$

Define the upper bound for $x_{in}(s)$ as ρ_{in}: $\rho_{in} := \|s(s-a)^{-1}\|_{\mathcal{L}_1}\rho_0$

Remark E.1.1.

The infinity-norm is defined for signals in time domain. As x_0 is a constant in Laplace domain, i.e. a Dirac delta function in time domain (whose $\mathcal{L}_\infty$-norm does not exist), it is expanded by 's' so that $\frac{x_0}{s}$ states a step signal in time domain for which the $\mathcal{L}_\infty$-norm exists.

Define an upper bound for x_{ref}: $\|x_{ref}\|_{\mathcal{L}_\infty} < \rho_r$

Define an upper bound for u_{ref}: $\|u_{ref}\|_{\mathcal{L}_\infty} < \rho_{ur}$

With a being Hurwitz and x_0 being bounded one has $\|x_{in}\|_{\mathcal{L}_\infty} \leq \rho_{in}$

Define a command prefilter constant $k_g := -(a^{-1}b)^{-1}$

The implemented low-pass filter $C(s)$ is one of the design parameters. The bandwidth can be chosen by the designer. It is subject to the constraints: strictly proper, stable, and DC-gain of $0\ dB$.

Assumption E.1.1.

An upper bound of sensor noise $\nu(iT) \in \mathbb{R}$ is given by $2\,|\nu|_{max} < \bar{\nu}$

Remark E.1.2.

It is necessary in Assumption E.1.1 to set a bound for the double maximum noise level as the measurement noise can jump from $-\,|\nu|_{max}$ to $|\nu|_{max}$, especially due to the high-pass transfer of noise to the prediction error (see Appendix J).

Measurement noise enters dynamic systems. The transfer functions $H_{\tilde{x}\nu,raw}(s)$ and $H_{\tilde{x}\nu,rec}(s)$ are shown later in Appendix J in (J.34) and Section J.4.

Define the following noise signals and their upper bounds.

$$\nu_{raw}(s) := H_{\tilde{x}\nu,raw}(s)\nu(s) + \varepsilon_{raw}(s); \qquad\qquad \bar{\nu}_{raw} \le \|H_{\tilde{x}\nu,raw}(s)\|_{\mathcal{L}_1}\bar{\nu} + \bar{\varepsilon}_{raw}$$

$$\tag{E.4}$$

$$\nu_{rec}(s) := H_{\tilde{x}\nu,rec}(s)\nu(s) + \varepsilon_{rec}(s); \qquad\qquad \bar{\nu}_{rec} \le \|H_{\tilde{x}\nu,rec}(s)\|_{\mathcal{L}_1}\bar{\nu} + \bar{\varepsilon}_{rec}$$

$$\tag{E.5}$$

$$\nu_h(s) := \frac{1}{Ts}H_{\tilde{x}\nu,rec}\nu(s) + \varepsilon_h(s) = H_{h\nu}(s)\nu(s) + \varepsilon_h(s); \qquad\qquad \bar{\nu}_h \le \|H_{h\nu}(s)\|_{\mathcal{L}_1}\bar{\nu} + \bar{\varepsilon}_h$$

$$\tag{E.6}$$

where $\bar{\varepsilon}_k$, $k \in \{raw, rec, h\}$ is the upper bound for the bounded approximation error ε_k (continuous - discrete) of the noise transfer functions. These occur since Appendix J shows a continuous approximation. The term $\frac{1}{Ts}$ in Equation (E.6) is later explained in Remark J.1.1, page 192.

Let $V(s)$ be defined by:

$$V(s) := \frac{C(s)\omega(s)}{1 + C(s)(\omega(s) - 1)} \tag{E.7}$$

Further define:

$$H_{xm}(s) := \frac{b}{s - a} \tag{E.8}$$

> For MIMO-systems:
> $$K_g H_m(s) := M(s) = K_g C(s\mathbb{I} - A)^{-1}B; \qquad H_{xm}(s) := (s\mathbb{I} - A)^{-1}B; \qquad \mathbb{I} \in \mathbb{R}^{n \times n}$$
> The DC-gain of $C(0)$ is $\mathbb{I}$.

Define κ, ϑ, Δ, ς:

$$\kappa := \int_0^T \left\| e^{a_{SP}(T-\tau)}b \right\|_\infty d\tau \tag{E.9}$$

$$\vartheta := \frac{\|(s - a)^{-1}bV(s)(s - a_{SP})/b\|_{\mathcal{L}_1}}{1 - \|(s - a)^{-1}b(1 - V(s))\|_{\mathcal{L}_1}L} \tag{E.10}$$

$$\Delta := \left(\left\| \frac{\omega - 1}{\omega} \right\|_\infty \|V(s)\|_{\mathcal{L}_1} + 1 \right) (L\rho_r + L_0) + \left\| \frac{\omega - 1}{\omega} \right\|_\infty \|V(s)\|_{\mathcal{L}_1} \left\| \frac{C_r(s)}{C(s)} k_g \right\|_{\mathcal{L}_1} \|r(s)\|_{\mathcal{L}_\infty}$$

$$+ \left(\left\| \frac{\omega - 1}{\omega} \right\|_\infty \|V(s)\|_{\mathcal{L}_1} \left(L\vartheta + \left\| \left(\frac{b}{s - a_{SP}} \right)^{-1} \right\|_{\mathcal{L}_1} \right) + L\vartheta \right) \bar{\gamma}_0 \tag{E.11}$$

$$\varsigma := \kappa \Delta + \bar{\nu}_{raw} \tag{E.12}$$

with $\bar{\gamma}_0 > 0$ being an arbitrarily small constant determining the required sampling time. It is chosen as $\bar{\gamma}_0 > \gamma_0$, where γ_0 is one of the performance bounds introduced later.

> For MIMO-systems:
> $$\kappa := \int_0^T \left\| e^{A_{SP}(T-\tau)} B \right\|_\infty d\tau \qquad \vartheta := \frac{\left\| (s\mathbb{I}-A)^{-1} BV(s)((s\mathbb{I}-A_{SP})^{-1}B)^{-1} \right\|_{\mathcal{L}_1}}{1 - \|(s\mathbb{I}-A)^{-1}B(\mathbb{I}-V(s))\|_{\mathcal{L}_1} L}$$
> $$\Delta := \left(\|\omega^{-1}(\omega - \mathbb{I})\|_\infty \|V(s)\|_{\mathcal{L}_1} + 1 \right) (L\rho_r + L_0)$$
> $$+ \|\omega^{-1}(\omega - \mathbb{I})\|_\infty \|V(s)\|_{\mathcal{L}_1} \|C(s)^{-1}C_r(s)K_g\|_{\mathcal{L}_1} \|r(s)\|_{\mathcal{L}_\infty}$$
> $$+ \left(\|\omega^{-1}(\omega - \mathbb{I})\|_\infty \|V(s)\|_{\mathcal{L}_1} \left(L\vartheta + \|H_{xm}^{-1}(s)\|_{\mathcal{L}_1} \right) + L\vartheta \right) \bar{\gamma}_0;$$

Assumption E.1.2.

The state of interest $x(t)$ is measurable.

Assumption E.1.3.

Of the unknown input gain $\omega(t) \in \mathbb{R} \setminus \{0\}$, $sgn(\omega)$ is known.

By merging the known or desired value of the input gain into b, the assumed input gain is set to $\omega_0 = 1$.

> For MIMO-systems:
>
> Assumption E.1.3 is expanded by: The unknown input gain matrix $\omega(t)$ is nonsingular and strictly row-diagonally-dominant.

Define X as $X = [x^T, z^T]^T$, then:

Assumption E.1.4.

There exists a $K_0 > 0$, such that:
$$\|f(t, X = 0)\|_\infty \leq K_0 \text{ for all } t \geq 0$$

Remark E.1.3.

The meaning of Assumption E.1.4 is that all disturbances which are not a function of $x(t)$ are bounded for all times.

Assumption E.1.5.

A Lipschitz constant $K > 0$ exists such that the following semiglobal Lipschitz condition holds:

$$\|f(t, X_2) - f(t, X_1)\|_{\mathcal{L}_\infty} \leq K \|X_2 - X_1\|_{\mathcal{L}_\infty}$$

Assumption E.1.6.

The z-dynamics are BIBO-stable, i.e. there exists a semiglobal Lipschitz constant $G > 0$ such that:

$$\|z_i\|_{\mathcal{L}_\infty} \leq G \|x_i\|_{\mathcal{L}_\infty} + G_0$$

Corollary E.1.1.

With assumption E.1.6 and E.1.5 it follows that a semiglobal Lipschitz constant $L > 0$ and an $L_0 > 0$ exist such that:

$$\|f(t, x_2, z(x_2)) - f(t, x_1, z(x_1))\|_{\mathcal{L}_\infty} \leq L \|x_2 - x_1\|_{\mathcal{L}_\infty} \text{ and}$$
$$\|f_i\|_{\mathcal{L}_\infty} \leq L \|x_i\|_{\mathcal{L}_\infty} + L_0$$

There exists a ρ_r such that $\rho_r > \rho_{in}$ (for a given ρ_0). With this, the following $\mathcal{L}_1$-norm condition must be satisfiable by the choice of $C(s)$:

$$\|H_{xm}(s)(1 - V(s))\|_{\mathcal{L}_1} < \frac{\rho_r - \rho_{in} - \left\|H_{xm}(s)V(s)\frac{C_r(s)}{C(s)}(s)k_g\right\|_{\mathcal{L}_1}\|r(s)\|_{\mathcal{L}_\infty}}{L\rho_r + L_0} \tag{E.13}$$

Remark E.1.4.

The $\mathcal{L}_1$-norm condition is necessary for the stability of the reference system later defined in section E.2 that assumes a perfect adaptive law. Hence, it is independent of the sampling time. The sampling time will be a dependence on the distance to the reference system.

Further define the following controller variables:

$$\Phi(t) := \int_0^t e^{a_{SP}(t-\tau)}d\tau = (-a_{SP})^{-1}\left(1 - e^{a_{SP}t}\right) \tag{E.14}$$

Define $\Upsilon(t)$ as:

$$\Upsilon(t) := e^{a_{SP}t} \tag{E.15}$$

Let the prediction error $\tilde{x}(t)$ be defined as $\tilde{x}(t) := \hat{x}(t) - x(t) - \nu(t)$, and define $\tilde{\sigma}(t) := \hat{\sigma}(t) - \sigma(t)$.

The adaptive law for updating the variable $\hat{\sigma}(t)$ is defined by:

$$\hat{\sigma}(t) = \hat{\sigma}(iT) := -b^{-1}\Phi(T)^{-1}\Upsilon(T)\tilde{x}(iT) \tag{E.16}$$

For MIMO-systems:
$$\Phi(t) := \int_0^t e^{A_{SP}(t-\tau)}d\tau = (-A_{SP})^{-1}\left(\mathbb{I} - e^{A_{SP}t}\right), \quad \mathbb{I} \in \mathbb{R}^{n \times n}; \quad \Upsilon(t) := e^{A_{SP}t}$$
$$\hat{\sigma}(iT) := -B^{-1}\Phi(T)^{-1}\Upsilon(T)\tilde{x}(iT)$$

With a_{SP} being a tuning parameter, define a state predictor as:

$$\dot{\hat{x}}(t) = ax(t) + b(\omega_0 u(t) + \hat{\sigma}(t)) + a_{SP}\tilde{x}(t), \quad \hat{x}(0) = x_0 + \nu(0) \tag{E.17}$$

The error-dynamics without measurement noise are defined as:

$$\dot{\tilde{x}}(t) = a_{SP}\tilde{x}(t) + b\tilde{\sigma}(t), \quad \tilde{x}(0) = 0 \tag{E.18}$$

The control law is defined as:

$$u(s) = C_r(s)k_g r(s) - C(s)\hat{\sigma}(s) \tag{E.19}$$

where $k_g = ((-a)^{-1}b)^{-1}$, so that the DC-gain of $H_{xm}(s)k_g = M(s)$ is one.

For MIMO-systems:
$$\dot{\hat{x}}(t) = Ax(t) + B(\omega_0 u(t) + \hat{\sigma}(t)) + A_{SP}\tilde{x}(t), \quad \hat{x}(0) = x_0 + \nu(0); \quad \hat{y}(t) = C\hat{x}(t)$$
$$\dot{\tilde{x}}(t) = A_{SP}\tilde{x}(t) + B\tilde{\sigma}(t), \quad \tilde{x}(0) = 0; \quad u(s) = C_r(s)K_g r(s) - C(s)\hat{\sigma}(s)$$
$$K_g = (C(-A)^{-1}B)^{-1}, \quad \text{DC-gain of diagonal elements of } M(s) \text{ is one.}$$

The command $r(s)$ can be used either as digital command as in a FBW-system, or as some direct mechanical input that is measured, or as baseline control signal. All kinds of command enter the input equation in the same way. See Chapter 3.1 for additional comments.

The bounds will be found as:

$$\gamma_0 := \max_{t\in[0,T]} \left(\int_0^t \left\| e^{a_{SP}(t-\tau)} b \right\|_\infty d\tau \right) \Delta + \bar{\nu}_{raw}, \quad \bar{\gamma}_0 > \gamma_0 \tag{E.20}$$

$$\gamma_1 := \vartheta \bar{\gamma}_0; \tag{E.21}$$

$$\gamma_2 := \left\| \frac{V(s)}{\omega(s)} \right\|_{\mathcal{L}_1} L\gamma_1 + \left\| \frac{V(s)\left(\frac{b}{s-a_{SP}}\right)^{-1}}{\omega(s)} \right\|_{\mathcal{L}_1} \bar{\gamma}_0 \tag{E.22}$$

E.2 Reference System

A reference system is established, which formulates a closed-loop plant while assuming a perfect adaptive law. The reference system has two inputs, namely the command $r(t)$ and internal dynamics output $z(t)$. The dynamic uncertainty $f(t, x_{ref}, z(t))$ is therefore expressed in terms of $x_{ref}(t)$ and $z(t)$. Hence, the output of the internal dynamics is not a disturbance for the reference system but rather the same $z(t)$ as for the plant. The purpose of introducing this reference system is to formulate stability first independent of the adaptive law. After formulating stability conditions of the reference system, an upper bound for the distance to the reference system is found that includes the adaptation quality in terms of the computational time step length T.

In contrast to the desired system however it includes the frequency limitations of control by the low-pass filter, i.e. a desired model that is supplemented by degradation of low-frequency control.

$$\dot{x}_{ref}(t) = a x_{ref}(t) + b\left(\omega(t)u_{ref}(t) + f(t, x_{ref}(t), z(t))\right) \tag{E.23}$$

with $x_{ref}(0) = x_0$

$$u_{ref}(s) = -\frac{V(s)}{\omega(s)}\eta_{ref}(s) + \frac{V(s)}{\omega(s)}\frac{C_r(s)}{C(s)}k_g r(s) \tag{E.24}$$

where $\eta_{ref}(s)$ is the Laplace transform of $f(t, x_{ref}(t), z(t))$.

Lemma E.2.1.

This reference system is BIBO-stable if the $\mathcal{L}_1$-norm condition (E.13) is fulfilled, and:

$$\|x_{ref}\|_{\mathcal{L}_\infty} < \rho_r, \qquad \|u_{ref}\|_{\mathcal{L}_\infty} < \rho_{ur}$$

For MIMO-systems:

$$\dot{x}_{ref}(t) = Ax_{ref}(t) + B\left(\omega(t)u_{ref}(t) + f(t, x_{ref}(t), z(t))\right); \qquad y_{ref}(t) = Cx_{ref}(t)$$

$$u_{ref}(s) = -\omega(s)^{-1}V(s)\eta_{ref}(s) + -\omega(s)^{-1}C(s)^{-1}V(s)C_r(s)K_g r(s)$$

The proof can be found in section E.4.1.

E.3 Core Proof

In this section, bounds for the error dynamics are derived. With these bounds, a bound for the distance to the reference system is found.

It is proved here that $\|\tilde{x}\|_{\mathcal{L}_\infty}$, $\|x_{ref} - x\|_{\mathcal{L}_\infty}$, $\|u_{ref} - u\|_{\mathcal{L}_\infty}$ are bounded by constants:

$$\|\tilde{x}\|_{\mathcal{L}_\infty} < \bar{\gamma}_0 \tag{E.25}$$

$$\|x_{ref} - x\|_{\mathcal{L}_\infty} < \gamma_1 \tag{E.26}$$

$$\|u_{ref} - u\|_{\mathcal{L}_\infty} < \gamma_2 \tag{E.27}$$

A time domain solution of the error dynamics (E.18) is computed. With $t_0 = iT$, $i = 0, 1, 2...$, and $t < T$, $\hat{\sigma}(iT)$ in the equation is seen as a constant. Adding sensor noise $\nu_{raw}(iT)$ delivers:

$$\tilde{x}(iT + t) = e^{a_{SP}t}\tilde{x}(iT) + \int_{iT}^{iT+t} e^{a_{SP}(iT+t-\tau)}b\hat{\sigma}(iT)d\tau \tag{E.28}$$

$$- \int_{iT}^{iT+t} e^{a_{SP}(iT+t-\tau)}b\sigma(\tau)d\tau + \nu_{raw}(iT)$$

This can be rewritten to:

$$\tilde{x}(iT + t) = e^{a_{SP}t}\tilde{x}(iT) + \int_0^t e^{a_{SP}(t-\tau)}b\hat{\sigma}(iT)d\tau \tag{E.29}$$

$$- \int_0^t e^{a_{SP}(t-\tau)}b\sigma(iT + \tau)d\tau + \nu_{raw}(iT)$$

Now split Equation (E.29) into two components for separating $\tilde{x}(iT)$ and $\hat{\sigma}(iT)$ from the rest, then:

$$\chi(iT + t) = e^{a_{SP}t}\tilde{x}(iT) + \int_0^t e^{a_{SP}(t-\tau)}b\hat{\sigma}(iT)d\tau \tag{E.30}$$

$$\zeta(iT + t) = - \int_0^t e^{a_{SP}(t-\tau)}b\sigma(iT + \tau)d\tau \tag{E.31}$$

The components are such that:

$$\tilde{x}(iT + t) = \chi(iT + t) + \zeta(iT + t) + \nu_{raw}(iT) \tag{E.32}$$

The upcoming proof of boundedness of $\tilde{x}(t)$ takes place in three steps.

First, boundedness is proved for $t = 0$.

Second, a next discrete step to $t = (i + 1)T$ is evaluated.

Third, all remaining times are evaluated by $\tilde{x}(iT + t)$.

Remark E.3.1.

An equation containing some dependency on iT describes a set of equations, one for every i. In the following, one not further specified i_0T is taken which is *one* choice of the set given by iT, according to $i = 0, 1, 2, \ldots$.

First, a measurement of the state is available, including measurement noise: $\tilde{x}(0) = 0$.

Second, $t = (i + 1)T$:

Applying $\hat{\sigma}(iT)$ shows that $\chi((i_0 + 1)T) = 0$. Then it follows with Equation (E.32):

$$\tilde{x}((i_0 + 1)T) = \zeta((i_0 + 1)T) + \nu_{raw}((i_0 + 1)T) \tag{E.33}$$

At this point, it becomes obvious that two mutually dependent bounds appear. In section E.4.2 we see that an upper bound for $\sigma(t)$, namely Δ, is dependent on an upper bound for $\tilde{x}(t)$, namely $\bar{\gamma}_0$. To avoid circular reasoning, the proof is conducted by contradiction. Starting in both (still to be proved) semiglobal positively invariant sets is assumed, since $\tilde{x}(0) = 0 \leq \varsigma$. Now we hypothetically go to the border of the set for $\tilde{x}(t)$.

The hypothesis to be refuted is: "$|\tilde{x}(t)| < \bar{\gamma}_0$ is not always true". By starting with $\tilde{x}(0) = 0$, a continuous $\tilde{x}(t)$ implies a time $0 < \iota$, where $|\tilde{x}(\iota)| = \bar{\gamma}_0$ and $|\tilde{x}(0 \leq t < \iota)| < \bar{\gamma}_0$.

This assumption is equivalent to the statement:

$$\|\tilde{x}_\iota\|_{\mathcal{L}_\infty} = \bar{\gamma}_0 \tag{E.34}$$

Lemma E.3.1.

$$\text{If } \|\tilde{x}_\iota\|_{\mathcal{L}_\infty} = \bar{\gamma}_0, \text{ then } \|\sigma_\iota\|_{\mathcal{L}_\infty} \leq \Delta$$

The proof can be found in section E.4.2.

Then:

$$\begin{aligned}
\|\tilde{x}((i_0 + 1)T)\|_\infty &= \|\zeta((i_0 + 1)T) + \nu_{raw}((i_0 + 1)T)\|_\infty \\
&\leq \int_0^T \left\|e^{a_{SP}(T-\tau)}b\right\|_\infty d\tau \, \|\sigma_\iota\|_{\mathcal{L}_\infty} + \bar{\nu}_{raw} \\
&= \kappa\Delta + \bar{\nu}_{raw} = \varsigma
\end{aligned} \tag{E.35}$$

It can be seen that $\lim\limits_{T \to 0} (\kappa\Delta) = 0$ as $\lim\limits_{T \to 0} \kappa = 0$ and Δ is bounded.

For MIMO-systems:
$$\zeta((i_0 + 1)T) \leq \int_0^T \left\|e^{A_{SP}(T-\tau)}B\right\|_\infty d\tau \, \|\sigma_\iota\|_{\mathcal{L}_\infty} + \bar{\nu}_{raw} = \kappa\Delta + \bar{\nu}_{raw} = \varsigma$$
where $\bar{\nu}_{raw}$ stems from a MIMO version of Appendix J.

This shows the main idea of the proof, namely that this term can be rendered arbitrarily small by reducing the sampling time T with measurement noise as the unavoidable remaining error.

Third, for additional times:

For: $0 < t \leq T$:

$$\|\zeta(i_0 T + t)\|_\infty \leq \max_{t \in [0,T]} \left(\int_0^t \left\|e^{a_{SP}(t-\tau)}b\right\|_\infty d\tau \right) \|\sigma_\iota\|_{\mathcal{L}_\infty} + \bar{\nu}_{raw} = \gamma_0 \tag{E.36}$$

As $\bar{\gamma}_0$ was chosen to be $\gamma_0 < \bar{\gamma}_0$, one has $\|\tilde{x}_i\|_\infty < \bar{\gamma}_0$.

Remark E.3.2.

In Equation (E.31) some terms show t, others T. To overcome assorted dependencies, the maximum in the span $t \in [0, T]$ is taken.

It can be seen that $\lim\limits_{T \to 0} \gamma_0 \leq \bar{\nu}_{raw}$ as $\lim\limits_{T \to 0} \kappa = 0$.

> For MIMO-systems:
> $$\zeta((i_0 T + t) \leq \int_0^t \left\| e^{A_{SP}(t-\tau)} B \right\|_\infty d\tau \, \|\sigma_i\|_{\mathcal{L}_\infty} + \bar{\nu}_{raw} = \gamma_0$$

Remark E.3.3.

For helicopters, considering times between iT and $(i+1)T$ is meaningless. The step length T is much shorter than relevant dynamics, hence, if the signals are bounded at iT and $(i+1)T$, they are as well bounded at times between and cannot change significantly.

This rounds off the proof via contradiction as the assumption in (E.34) is refuted by this.

The value of $\bar{\gamma}_0$ was chosen beforehand as some guessed value and first appears in (E.34). For completing the contradiction, it must be true that $\tilde{x}(iT + t) < \bar{\gamma}_0$ which implies that $\gamma_0 < \bar{\gamma}_0$ has to hold. With $\lim\limits_{T \to 0} \kappa = 0$ it is clear that for every $\bar{\gamma}_0$ a limit of $\tilde{x}(iT + t)$, namely γ_0 can be found, if only the computation step length T is chosen small enough.

By this method, circular reasoning is avoided. This would occur if T is chosen, and the performance bounds are tried to be found. By iterating this procedure - choosing $\bar{\gamma}_0$ and finding the necessary maximum T, the design can be theoretically done by first stating the performance bounds $\bar{\gamma}_0$ and finding the required CPU performance. With this, it must not be forgotten that the bounds are done with very conservative elements in the proof and are therefore not verified for most applications.

Next, $\|(x_{ref} - x)_i\|_{\mathcal{L}_\infty}$ and $\|(u_{ref} - u)_i\|_{\mathcal{L}_\infty}$ are analyzed:

Lemma E.3.2.

$$\text{For } \|\tilde{x}_i\|_{\mathcal{L}_\infty} < \bar{\gamma}_0 \text{ one has } \|(x_{ref} - x)_i\|_{\mathcal{L}_\infty} < \vartheta \bar{\gamma}_0.$$

The proof can be found in section E.4.3.

Then:

$$\|e_l\|_{\mathcal{L}_\infty} = \|(x_{ref} - x)_l\|_{\mathcal{L}_\infty} < \vartheta\bar{\gamma}_0 = \gamma_1 \tag{E.37}$$

Likewise, $\|(u_{ref} - u)_l\|_{\mathcal{L}_\infty}$ is bounded by a constant γ_2:

Lemma E.3.3.

$$\|(u_{ref} - u)_l\|_{\mathcal{L}_\infty} < \left\|\frac{V(s)}{\omega(s)}\right\|_{\mathcal{L}_1} L\gamma_1 + \left\|\frac{V(s)\left(\frac{b}{s-a_{SP}}\right)^{-1}}{\omega(s)}\right\|_{\mathcal{L}_1} \bar{\gamma}_0 = \gamma_2$$

The proof can be found in section E.4.4.

This means that a time l is not reached and the bounds hold for all times. Thus (E.25), (E.26), and (E.27) are proved.

Remark E.3.4.

The non-implemented low-pass filter $V(s)$ comprises the uncertain input gain $\omega(s)$. Thus, the $\mathcal{L}_1$-norm condition contains uncertainties not only in output-feedback. There, this uncertainty is – among others – included in $H(s)$.

Summary of inclusion of the measurement noise:

By injecting measurement noise into the predictor, dynamic effects occur, namely the current noise ratio in $\tilde{x}(iT)$ is dependent on the current noise and its entire history. Approximating discrete behavior by continuous transfer functions allows for a correct evaluation of low-frequency noise, high frequencies however are not correctly described by it. For the latter, the current addition of every time step dominates the effects from history, the noise enters the prediction error through a virtual discrete high-pass filter. Hence, within the time span of $t \in [iT, (i+1)T]$ a more accurate description is the addition of new noise $\nu((i+1)T)$ every $t = (i+1)T$, meaning that $\nu(iT)$ is somehow included in the approximation error $\varepsilon(t)$.

The measurement noise is primarily a degradation of the adaption quality expressed in $\tilde{x}(t)$, which is then propagated to the errors $x_{ref}(t) - x(t)$ and $u_{ref}(t) - u(t)$.

E.4 Proof of Lemmas

E.4.1 Proof of Lemma E.2.1

$$\dot{x}_{ref}(t) = ax_{ref}(t) + b(\omega(t)u_{ref}(t) + f(t, x_{ref}(t), z(t))) \tag{E.38}$$

With $\eta_{ref}(s)$ being the Laplace transformation of $f(t, x_{ref}(t), z(t))$, one has from Equation (E.24):

$$u_{ref}(s) = -\frac{V(s)}{\omega(s)}\eta_{ref}(s) + \frac{V(s)}{\omega(s)}\frac{C_r(s)}{C(s)}k_g r(s) \tag{E.39}$$

Thus:

$$x_{ref}(s) = \underbrace{\frac{b}{s-a}}_{H_{xm}(s)}\left((1 - V(s))\eta_{ref}(s) + V(s)\frac{C_r(s)}{C(s)}k_g r(s)\right) + x_{in}(s) \tag{E.40}$$

It is proceeded by contradiction. If the bound $\|x_{ref}\|_{\mathcal{L}_\infty} < \rho_r$ does not hold but is true in $t = 0$, it means that there exists a $t = \imath$ such that $\|x_{ref,\imath}\|_{\mathcal{L}_\infty} = \rho_r$:

With $\|f_\imath\|_{\mathcal{L}_\infty} \leq L\rho_r + L_0$:

$$\|x_{ref,\imath}\|_{\mathcal{L}_\infty} \leq \|H_{xm}(s)(1 - V(s))\|_{\mathcal{L}_1}(L\rho_r + L_0) + \left\|H_{xm}(s)V(s)\frac{C_r(s)}{C(s)}(s)k_g\right\|_{\mathcal{L}_1}\|r\|_{\mathcal{L}_\infty} + \rho_{in} \tag{E.41}$$

The $\mathcal{L}_1$-norm condition of (E.13) can be solved for ρ_r :

$$\|H_{xm}(s)(1 - V(s))\|_{\mathcal{L}_1}(L\rho_r + L_0) + \left\|H_{xm}(s)V(s)\frac{C_r(s)}{C(s)}(s)k_g\right\|_{\mathcal{L}_1}\|r\|_{\mathcal{L}_\infty} + \rho_{in} < \rho_r$$

showing that $\|x_{ref,\imath}\|_{\mathcal{L}_\infty} < \rho_r$, thus contradicting the above mentioned assumption ("the bound $\|x_{ref}\|_{\mathcal{L}_\infty} < \rho_r$ does not hold"). With this, $\|x_{ref}\|_{\mathcal{L}_\infty} < \rho_r$ is proved. Furthermore, it is confirmed that $\|f_\imath\|_{\mathcal{L}_\infty} \leq L\rho_r + L_0$.

Remark E.4.1.
An intuitive explanation is shown on page 27 with Equation (2.12) and Equation (2.13), why the $\mathcal{L}_1$-norm condition benefits from high low-pass bandwidth.

With this result, one also has:

$$\|u_{ref,i}\|_{\mathcal{L}_\infty} < \left\|\frac{V(s)}{\omega(s)}\right\|_{\mathcal{L}_1} (L\rho_r + L_0) + \left\|\frac{V(s)}{\omega(s)}\frac{C_r(s)}{C(s)}k_g\right\|_{\mathcal{L}_1} \|r\|_{\mathcal{L}_\infty} = \rho_{ur} \tag{E.42}$$

$\square$

E.4.2 Proof of Lemma E.3.1

Considering the plant input (E.19), one has:

$$u(s) = -C(s)\hat\sigma(s) + C_r(s)k_g r(s) \tag{E.43}$$

where $\hat\sigma(s)$ is the Laplace transform of $\hat\sigma(t) = \hat\sigma(iT)$, and:

$$\sigma(t) = (\omega(t) - \omega_0)u(t) + f(t, x(t), z(t)) \tag{E.44}$$

With $\omega_0 = 1$, $\hat\sigma(t) = \tilde\sigma(t) + \sigma(t)$ and Equation (E.44):

$$u(s) = -C(s)\hat\sigma(s) + C_r(s)k_g r(s) = -C(s)\left((\omega(t) - 1)u(s) + \eta(s) + \tilde\sigma(s)\right) + C_r(s)k_g r(s) \tag{E.45}$$

$$u(s) = \frac{-C(s)}{1 + C(s)(\omega(s) - 1)}(\eta(s) + \tilde\sigma(s)) + \frac{C_r(s)}{1 + C(s)(\omega(s) - 1)}k_g r(s) \tag{E.46}$$

$$\frac{-C(s)}{1 + C(s)(\omega(s) - 1)} = -\frac{V(s)}{\omega(s)} \tag{E.47}$$

Consequently, (E.46) rewritten is:

$$\omega(s)u(s) = -V(s)\left(\eta(s) + \tilde\sigma(s)\right) + V(s)\frac{C_r(s)}{C(s)}k_g r(s) \tag{E.48}$$

Remark E.4.2.

This calculation is consistent with the expression from the reference system in Equation (E.24) and clarifies its meaning.

By merging (E.48) and (E.44) one has:

$$\sigma(s) = -\frac{\omega(s) - 1}{\omega(s)} V(s) \left(\eta(s) + \tilde{\sigma}(s) - \frac{C_r(s)}{C(s)} k_g r(s) \right) + \eta(s) \tag{E.49}$$

Corollary E.1.1 implies:

$$f(t, x(t), z(t)) \leq L \left\| x_l \right\|_{\mathcal{L}_\infty} + L_0 \tag{E.50}$$

From Lemma E.3.2:

$$\left\| x_l \right\|_{\mathcal{L}_\infty} \leq \rho_r + \vartheta \left\| \tilde{x}_l \right\|_{\mathcal{L}_\infty} \tag{E.51}$$

It follows that:

$$\left\| \eta \right\|_{\mathcal{L}_\infty} \leq L \left(\rho_r + \vartheta \left\| \tilde{x}_l \right\|_{\mathcal{L}_\infty} \right) + L_0 \tag{E.52}$$

Then, with

$$\tilde{\sigma}(s) = \left(\frac{b}{s - a_{SP}} \right)^{-1} \tilde{x}(s) \tag{E.53}$$

we have:

$$\begin{aligned}
\left\| \sigma_l \right\|_{\mathcal{L}_\infty} & \\
\leq \left\| \frac{\omega - 1}{\omega} \right\|_\infty & \left\| V(s) \right\|_{\mathcal{L}_1} \left(L\rho_r + L\vartheta \left\| \tilde{x}_l \right\|_{\mathcal{L}_\infty} + L_0 + \left\| \left(\frac{b}{s - a_{SP}} \right)^{-1} \right\|_{\mathcal{L}_1} \left\| \tilde{x}_l \right\|_{\mathcal{L}_\infty} + \left\| \frac{C_r(s)}{C(s)} k_g \right\|_{\mathcal{L}_1} \left\| r(s) \right\|_{\mathcal{L}_\infty} \right) \\
& + L\rho_r + L\vartheta \left\| \tilde{x}_l \right\|_{\mathcal{L}_\infty} + L_0
\end{aligned} \tag{E.54}$$

$$\begin{aligned}
\left\| \sigma_l \right\|_{\mathcal{L}_\infty} \leq & \left(\left\| \frac{\omega - 1}{\omega} \right\|_\infty \left\| V(s) \right\|_{\mathcal{L}_1} + 1 \right) \left(L\rho_r + L_0 \right) \\
& + \left\| \frac{\omega - 1}{\omega} \right\|_\infty \left\| V(s) \right\|_{\mathcal{L}_1} \left\| \frac{C_r(s)}{C(s)} k_g \right\|_{\mathcal{L}_1} \left\| r(s) \right\|_{\mathcal{L}_\infty} \\
& + \left(\left\| \frac{\omega - 1}{\omega} \right\|_\infty \left\| V(s) \right\|_{\mathcal{L}_1} \left(L\vartheta + \left\| \left(\frac{b}{s - a_{SP}} \right)^{-1} \right\|_{\mathcal{L}_1} \right) + L\vartheta \right) \left\| \tilde{x}_l \right\|_{\mathcal{L}_\infty}
\end{aligned} \tag{E.55}$$

If $\left\| \tilde{x}_l \right\|_{\mathcal{L}_\infty} = \bar{\gamma}_0$, then $\left\| \sigma_l \right\|_{\mathcal{L}_\infty} \leq \Delta$

$\square$

E.4.3 Proof of Lemma E.3.2

The plant in Laplace domain is:

$$x(s) = (s-a)^{-1}b(u(s) + \sigma(s)) + x_{in}(s) \tag{E.56}$$

Substituting (E.49) and (E.48) into the plant equation, it follows the closed-loop plant:

$$x(s) = (s-a)^{-1}b\left(-\frac{V(s)}{\omega}(\eta(s) + \tilde{\sigma}(s)) - \frac{\omega-1}{\omega}V(s)(\eta(s) + \tilde{\sigma}(s)) + \eta(s)\right) \tag{E.57}$$

$$x(s) = (s-a)^{-1}b\left((1 - V(s))\eta(s) - V(s)\tilde{\sigma}(s)\right) + x_{in}(s) \tag{E.58}$$

and:

$$x_{ref}(s) = (s-a)^{-1}b\left(\omega(s)u_{ref}(s) + \eta_{ref}(s)\right) + x_{ref,in}(s) \tag{E.59}$$

where $x_{ref,in}(s) = x_{in}(s)$

The reference input without any command is:

$$u_{ref}(s) = -\frac{V(s)}{\omega(s)}\eta_{ref}(s) \tag{E.60}$$

Then:

$$x_{ref}(s) = (s-a)^{-1}b\left((1 - V(s))\eta_{ref}(s)\right) + x_{ref,in}(s) \tag{E.61}$$

$$x_{ref}(s) - x(s) = (s-a)^{-1}b(1 - V(s))(\eta_{ref}(s) - \eta(s)) + (s-a)^{-1}bV(s)\tilde{\sigma}(s) \tag{E.62}$$

With Corollary E.1.1 one finds that:

$$\left\|(\eta_{ref} - \eta)_t\right\|_{\mathcal{L}_\infty} \leq L\left\|(x_{ref} - x)_t\right\|_{\mathcal{L}_\infty} \tag{E.63}$$

The error dynamics from Equation (E.19) are:

$$\dot{\tilde{x}}(t) = a_{SP}\tilde{x}(t) + b\tilde{\sigma}(t) \tag{E.64}$$

Their Laplace transformation are:

$$\tilde{x}(s) = (s - a_{SP})^{-1} b\tilde{\sigma}(s); \quad \tilde{x}_{in}(s) = 0 \tag{E.65}$$

Then it follows with Equation (E.65):

$$\tilde{\sigma}(s) = \frac{s - a_{SP}}{b}\tilde{x}(s) \tag{E.66}$$

Connecting both facts with Equation (E.62), delivers:

$$\left\|(x_{ref} - x)_l\right\|_{\mathcal{L}_\infty} \leq \left\|(s - a)^{-1}b(1 - V(s))\right\|_{\mathcal{L}_1} L \left\|(x_{ref} - x)_l\right\|_{\mathcal{L}_\infty} \tag{E.67}$$
$$+ \left\|(s - a)^{-1}bV(s)(s - a_{SP})/b\right\|_{\mathcal{L}_1} \left\|\tilde{x}_l\right\|_{\mathcal{L}_\infty}$$

$$\left\|(x_{ref} - x)_l\right\|_{\mathcal{L}_\infty} \leq \frac{\left\|(s - a)^{-1}bV(s)(s - a_{SP})/b\right\|_{\mathcal{L}_1}}{1 - \left\|(s - a)^{-1}b(1 - V(s))\right\|_{\mathcal{L}_1} L} \left\|\tilde{x}_l\right\|_{\mathcal{L}_\infty} = \vartheta \left\|\tilde{x}_l\right\|_{\mathcal{L}_\infty} \tag{E.68}$$

for MIMO-sytems:
$$\left\|(x_{ref} - x)_l\right\|_{\mathcal{L}_\infty} \leq \frac{\left\|(s\mathbb{I}-A)^{-1}BV(s)((s\mathbb{I}-A_{SP})^{-1}B)^{-1}\right\|_{\mathcal{L}_1}}{1-\left\|(s\mathbb{I}-A)^{-1}B(\mathbb{I}-V(s))\right\|_{\mathcal{L}_1}L} \left\|\tilde{x}_l\right\|_{\mathcal{L}_\infty} = \vartheta \left\|\tilde{x}_l\right\|_{\mathcal{L}_\infty}$$

Note that for $\left\|\tilde{x}_l\right\|_{\mathcal{L}_\infty} < \bar{\gamma}_0$ one has $\left\|(x_{ref} - x)_l\right\|_{\mathcal{L}_\infty} < \vartheta\bar{\gamma}_0 = \gamma_1$.

$\square$

E.4.4 Proof of Lemma E.3.3

From Equation (E.39) and (E.48) it follows that:

$$u_{ref}(s) = -\frac{V(s)}{\omega(s)}\eta_{ref}(s) + \frac{V(s)}{\omega(s)}\frac{C_r(s)}{C(s)}k_g r(s) \tag{E.69}$$

$$u(s) = -\frac{V(s)}{\omega(s)}(\eta(s) + \tilde{\sigma}(s)) + \frac{V(s)}{\omega(s)}\frac{C_r(s)}{C(s)}k_g r(s) \tag{E.70}$$

$$u_{ref}(s) - u(s) = \frac{V(s)}{\omega(s)}(\tilde{\sigma}(s) + \eta(s) - \eta_{ref}(s)) \tag{E.71}$$

with Corollary E.1.1 and the section above one has

$$\|(\eta - \eta_{ref})_i\|_{\mathcal{L}_\infty} \leq L \|(x - x_{ref})_i\|_{\mathcal{L}_\infty} < L\gamma_1.$$

With $\tilde{\sigma}(s) = \left(\frac{b}{s-a_{SP}}\right)^{-1} \tilde{x}(s)$ from (E.66) it follows that:

$$\|(u_{ref}(s) - u(s))_i\|_{\mathcal{L}_\infty} < \left\|\frac{V(s)}{\omega(s)}\right\|_{\mathcal{L}_1} L\gamma_1 + \left\|\frac{V(s)\left(\frac{b}{s-a_{SP}}\right)^{-1}}{\omega(s)}\right\|_{\mathcal{L}_1} \bar{\gamma}_0 = \gamma_2 \tag{E.72}$$

This confirms Lemma E.3.3 as $V(s)$ and $H_{xm}(s)$ are strictly proper and stable.

$\square$

E.5 Alternative, Recursive Adaptive Law

This modified adaptive law, introduced in [18] for achieving smaller values of $|\tilde{x}(t)|$, adds a recursive term to the adaptive law. By improving the quality of the prediction error, a smaller γ_0 propagates through to γ_1, i.e. a smaller deviation from the reference system. Therefore, only Appendix E.3 needs to be adjusted towards new definitions of γ_0 and $\bar{\gamma}_0$.

Define $\gamma_{0,z}$ as a replacement for γ_0:

$$\gamma_{0,z} = \left(\max_{t\in[0,T]}\left(e^{a_{SP}t}\right) + \max_{t\in[0,T]}\left(\int_0^t e^{a_{SP}(t-\tau)}d\tau\right)\Phi^{-1}(T)\Upsilon(T)\right)2\kappa\Delta \tag{E.73}$$
$$+ \max_{t\in[0,T]}\left(\int_0^t e^{a_{SP}(t-\tau)}bd\tau\right)\left(\kappa\Phi^{-1}(T) - b\right)2\Delta + \bar{\nu}_{rec}$$

The adaptive law for updating the variable $\hat{\sigma}(t)$ is defined as:

$$\hat{\sigma}(t) = \hat{\sigma}(iT) = -b^{-1}\Phi^{-1}(T)\Upsilon(T)\tilde{x}(iT) + b^{-1}\Phi^{-1}(T)h(iT), \quad i = 0, 1, 2, \dots \tag{E.74}$$

where:

$$h(iT) = -\tilde{x}(iT) + h((i-1)T), \quad h(0) = 0, \ i = 1, 2, \dots \tag{E.75}$$

meaning that the recursion starts with $h(0) = 0$ and continues with $h(1T) = -\tilde{x}(1T)$.

In an adaptive law without the additional term with $h(iT)$, Equation (E.31) integrated in the span $t \in [0, T]$ states the remaining term, i.e. $R = -\int_0^T e^{a_{SP}(T-\tau)} b\sigma(iT + \tau)d\tau$. It arises in the time span in which the old cumulated uncertainties are canceled while new uncertainties are cumulating again. See section 2.5.2 for a more precise explanation. With the recursive definition of $h(iT)$ based on $\tilde{x}(iT)$, it can be seen that these remaining terms are taken into account via $h(iT)$. Then, $h(iT)$ can be written as:

$$h(iT) = \int_0^T e^{a_{SP}(T-\tau)} b\sigma((i-1)T + \tau)d\tau + \nu_h(iT) \tag{E.76}$$

Consequently

$$\tilde{x}((i_0 + 1)T) = \int_0^T e^{a_{SP}(T-\tau)} b\sigma((i_0 - 1)T + \tau)d\tau \tag{E.77}$$
$$- \int_0^T e^{a_{SP}(T-\tau)} b\sigma(i_0 T + \tau)d\tau + \nu_{rec}((i_0 + 1)T)$$

These equations state the closed loop behavior, i.e. with $\hat{\sigma}$ being applied.

Proceeding with the proof by contradiction with Lemma E.3.1, i.e. $\sigma \leq \Delta$, an upper bound can be derived with $\|\tilde{x}_l\|_{\mathcal{L}_\infty}$:

$$\|\tilde{x}((i_0 + 1)T)\|_\infty \leq 2\kappa\Delta + \bar{\nu}_{rec} \tag{E.78}$$

Remark E.5.1.
In [18] an additional, theoretically smaller bound is shown. This is omitted here for simplicity reasons. The theoretical bounds cannot be verified due to lack of system knowledge and do not change the architecture.

Next is the proof of $\tilde{x}(iT + t) \leq \gamma_{0,z}$, where $\gamma_{0,z}$ is a constant and $0 \leq t \leq T$, $t \leq \iota - T$. From the initialization we have $\tilde{x}(t) = 0$. It follows from boundedness at $t = i_0 T$ that the bounds hold for $t = (i_0 + 1)T$, too. With the addition of the times $t \in [iT, (i+1)T \leq \iota)$, these facts provide a bound for all times t. It is shown that a time $t = \iota$ where $\tilde{x}(\iota) = \bar{\gamma}_{0,z}$ is not reached.

Merging $\hat{\sigma}(i_0 T)$ from (E.74) into (E.29), one has with $t < T$:

$$\tilde{x}(iT + t) = e^{a_{SP}t}\tilde{x}(iT)$$
$$+ \int_0^t e^{a_{SP}(t-\tau)} b \left(-b^{-1}\Phi^{-1}(T)\Upsilon(T)\tilde{x}(iT) + b^{-1}\Phi^{-1}(T)(h(iT) - \nu_h(iT))\right) d\tau$$
$$- \int_0^t e^{a_{SP}(t-\tau)} b\sigma(iT + \tau)d\tau + \nu_{rec}(iT)$$

$$(E.79)$$

This is equivalent to:

$$\tilde{x}(iT + t) = e^{a_{SP}t}\tilde{x}(iT) + \int_0^t e^{a_{SP}(t-\tau)} b(-b^{-1})\Phi^{-1}(T)\Upsilon(T)\tilde{x}(iT)d\tau$$
$$+ \underbrace{\int_0^t e^{a_{SP}(t-\tau)} bb^{-1}\Phi^{-1}(T)(h(iT) - \nu_h(iT))d\tau - \int_0^t e^{a_{SP}(t-\tau)} b\sigma(iT + \tau)d\tau}_{\varpi}$$
$$+\nu_{rec}(iT)$$

$$(E.80)$$

Note that a noise-free $h(iT)$, namely $h(iT) - \nu_h(iT)$ is written to uniquely define the noise effects by $\nu_{rec}(iT)$.

By substituting $h(iT)$ from Equation (E.76), and with the definition of $\Phi(t) = \int_0^t e^{a_{SP}(t-\tau)}d\tau = \Phi(T)$ from (E.14) one has:

$$\varpi = \Phi(t)bb^{-1}\Phi^{-1}(T) \underbrace{\int_0^T e^{a_{SP}(T-\tau)} b\sigma((i-1)T + \tau)d\tau}_{h(iT)-\nu(iT)} - \int_0^t e^{a_{SP}(t-\tau)} b\sigma(iT + \tau)d\tau \quad (E.81)$$

The expressions can be expanded to get:

$$\varpi = \Phi(t)bb^{-1}\Phi^{-1}(T) \int_0^T e^{a_{SP}(T-\tau)}b \underbrace{\sigma((i-1)T + \tau)}_{\sigma((i-1)T+\tau)-\sigma(iT)+\sigma(iT)} d\tau$$
$$- \int_0^t e^{a_{SP}(t-\tau)}b \underbrace{\sigma(iT + \tau)}_{\sigma(iT+\tau)-\sigma(iT)+\sigma(iT)} d\tau$$

$$(E.82)$$

with $\int_0^T e^{a_{SP}(T-\tau)}bd\tau = \Phi(T)$, one has:

$$\varpi = \Phi(t)\Phi^{-1}(T) \int_0^T e^{a_{SP}(T-\tau)}b \left(\sigma((i-1)T + \tau) - \sigma(iT)\right) d\tau$$
$$- \int_0^t e^{a_{SP}(t-\tau)}b \left(\sigma(iT + \tau) - \sigma(iT)\right) d\tau$$

$$(E.83)$$

Then an upper bound for (E.80) is found:

$$\|\tilde{x}(iT+t)\|_\infty \leq \left(\max_{t\in[0,T]} \left(e^{a_{SP}t} \right) + \max_{t\in[0,T]} \left(\int_0^t e^{a_{SP}(t-\tau)} d\tau \right) \Phi^{-1}(T)\Upsilon(T) \right) \kappa\Delta + \bar{\varpi} + \bar{\nu}_{rec}$$
$$= \gamma_{0,z}$$

(E.84)

where

$$\bar{\varpi} = \max_{t\in[0,T]} \left(\int_0^t e^{a_{SP}(t-\tau)} d\tau \right) \kappa\Phi^{-1}(T)2\Delta - \max_{t\in[0,T]} \left(\int_0^t e^{a_{SP}(t-\tau)} b\,d\tau \right) 2\Delta \quad \text{(E.85)}$$
$$= \max_{t\in[0,T]} \left(\int_0^t e^{a_{SP}(t-\tau)} d\tau \right) \left(\kappa\Phi^{-1}(T) - b \right) 2\Delta$$

for MIMO-sytems:
$$\gamma_{0,z} = \left(\max_{t\in[0,T]} \left(e^{A_{SP}t} \right) + \max_{t\in[0,T]} \left(\int_0^t e^{A_{SP}(t-\tau)} dB\tau \right) \Phi^{-1}(T)\Upsilon(T) \right) 2\kappa\Delta$$
$$+ \max_{t\in[0,T]} \left(\int_0^t e^{A_{SP}(t-\tau)} d\tau \right) \left(\kappa\Phi^{-1}(T) - B \right) 2\Delta + \bar{\nu}_{rec}$$

Remark E.5.2.

In Equation (E.80) some terms show t, others T. To overcome assorted dependencies, the maximum in the span $t \in [0,T]$ is taken.

This completes the proof of contradiction as $\bar{\gamma}_{0,z}$ is chosen such that $\gamma_{0,z} < \bar{\gamma}_{0,z}$. The new $\gamma_{0,z}$ and $\bar{\gamma}_{0,z}$ replace in case of application of the modified adaptive law γ_0 from Equation (E.36) and $\bar{\gamma}_0$.

Remark E.5.3.

In [18] a theoretically smaller performance bound is shown by exploiting the change rate of $\sigma(t)$. This is omitted here for simplicity reasons. The theoretical bounds cannot be verified due to lack of system knowledge and do not change the architecture.

Appendix F

Performance Bounds in Output Feedback for Non-SPR Desired Dynamics with a Recursive Adaptive Law

The ideas of the subsequent proof are based on [13], Chapter 4.2. This proof is first shown – in a much simplified version however – in [42]. The reader however is not required to be familiar with this book – the following proof tries to be complete. Modifications and additional explanations however apply.

The changes to the proof in [13] comprise:
The recursive adaptive law introduced in [18] for state feedback is applied here; an initialization procedure shown in [33] is included, that relaxes restrictions on the initial condition; conditions on the relative degree of $M(s)$ are examined; sensor noise is integrated in this proof; the feedforward filter $C_r(s)$ for $r(s)$ is introduced; the choice of $C(s)$ is based on hardware considerations instead of robust stability of $H(s)$, which on the contrary is checked for robustness against parametric uncertainties;

The basic ideas of the proof can be condensed in the following steps:
An unknown SISO transfer function affected by time-varying uncertainties and external disturbances is to be controlled. The uncertainties for this transfer function are modeled as scalar variable. An output predictor with vectorial uncertainty (thus in general unmatched, the term "unmatched uncertainty" is introduced in Definition A.1.14), an adaptive law and a control law with low-pass characteristics are the elements of the controller. For proofing stability, the error dynamics are transformed so that the output is the first component of

the transformed unknown error states. To the output error, a modified piece-wise constant adaptive law is applied. Boundedness of the error in internal states is proved by a positive definite function, similar to Lyapunov functions. In the first segment, stability is proved in broken-loop mode. Then, the adaptive control signal is added. This proof is done by contradiction due to two mutually dependent semiglobal positively invariant sets. Moving hypothetically to the bound of one of the sets results in the proof of the other invariant set to be true and the exclusion of the bound of the former, which again proofs the original set and contradicts the assumption of the performance bound ever reaching the bound itself. The reference system serves as intermediate step for the deviation of the desired from the real dynamics.

F.1 Definitions, Assumptions, Descriptions

A discrete controller is applied to a continuous plant. The computation rate, i.e. time step length of the algorithm, that is hard-coded in the implementation is denoted as T. It is subject to hardware limitations and chosen as low as task execution time allows.

The open-loop plant dynamics $F(s)$ are modeled as SISO transfer function per axis of unknown order and relative degree, affected by disturbances and output dependent nonlinearities $d(s)$. This is a short denotation for $d(t) = d'(t, y(t))$ where $d'(t, y(t))$ is an unknown, in general nonlinear map, $d' : \mathbb{R} \times \mathbb{R} \to \mathbb{R}$.

$$y(s) = F(s)(u(s) + d(s)) \tag{F.1}$$

Assumption F.1.1.
The lower bound of $reldeg(F(s))$ is known to be 1.

Remark F.1.1.
As substantiated in section 3.3, state and output predictors are chosen for each axis separately without cross-couplings, thus a SISO controller for each axis is permissible.

Assumption F.1.2.
In general, only the output $y(t)$ is measurable, the vector of states $x(t)$ is not.

Assumption F.1.3.
An upper bound for sensor noise $\nu(iT) \in \mathbb{R}$ affecting the system output $y(t)$ is given by $2\,|\nu|_{max} < \bar{\nu}$.

Remark F.1.2.

It is necessary in Assumption F.1.3 to set a bound for the double maximum noise level as the measurement noise can jump from $-|\nu|_{max}$ to $|\nu|_{max}$, especially due to the high-pass transfer of noise to the prediction error (see Appendix J).

Measurement noise enters dynamic systems. The transfer function $H_{\tilde{y}\nu,rec}(s)$ is shown later in Equation (J.39), $H_{\tilde{x}\nu}(s)$ is defined in (J.43).

Define the following noise signals and their upper bounds.

$$\nu_{rec}(s) := H_{\tilde{y}\nu,rec}(s)\nu(s) + \varepsilon_{rec}(s); \qquad\qquad \bar{\nu}_{rec} \leq \|H_{\tilde{y}\nu,rec}(s)\|_{\mathcal{L}_1} \bar{\nu} + \bar{\varepsilon}_{rec}$$

$$\text{(F.2)}$$

$$\nu_h(s) := \frac{1}{Ts}H_{\tilde{y}\nu,rec}\nu(s) + \varepsilon_h(s) = H_{h\nu}(s)\nu(s) + \varepsilon_h(s); \qquad \bar{\nu}_h \leq \|H_{h\nu}(s)\|_{\mathcal{L}_1} \bar{\nu} + \bar{\varepsilon}_h$$

$$\text{(F.3)}$$

$$\nu_{\tilde{x}}(s) := H_{\tilde{x}\nu}(s)\nu(s) + \varepsilon_{\tilde{x}}(s); \qquad\qquad \bar{\nu}_{rec} \leq \|H_{\tilde{y}\nu,rec}(s)\|_{\mathcal{L}_1} \bar{\nu} + \bar{\varepsilon}_{\tilde{x}}$$

$$\text{(F.4)}$$

where $\bar{\varepsilon}_k$, $k \in \{rec, h, \tilde{x}\}$ is the upper bound for the bounded approximation error ε_k (continuous - discrete) of the noise transfer functions. These occur since Appendix J shows a continuous approximation. The term $\frac{1}{Ts}$ in Equation (F.3) is later explained in Remark J.1.1.

Assumption F.1.4.

The map $d'(t, y(t))$ is subject to Lipschitz continuity, where L and L_0 exist for $t \geq 0$:
$$|d'(t, y_2) - d'(t, y_1)| \leq L\,|y_2 - y_1|\,; \qquad |d'(t, y)| \leq L\,|y| + L_0$$

The desired transfer function $M(s)$ which is stable, strictly proper, minimum phase (neither poles nor zeros in 0 or positive), states feasible (or almost feasible) dynamics. The minimal state space representation of $M(s)$ is given by (A, b, c^T). For its application to helicopters, $M(s)$ must be chosen to have relative degree 1. See section F.5 on page 162 for an explanation.

Rewriting the SISO plant dynamics (F.1) in terms of desired dynamics gives:

$$y(s) = M(s)(u(s) + \sigma(s)) \qquad\qquad \text{(F.5)}$$

All disturbances together with undesired dynamics are lumped into one scalar variable $\sigma(s)$, so that Equation (F.5) is equivalent to Equation (F.1), with $\sigma(s)$ containing all differences and disturbances.

$$\sigma(s) = \frac{(F(s) - M(s))u(s) + F(s)d(s)}{M(s)} \tag{F.6}$$

The plant (F.5) in time domain as minimal state space representation is:

$$\dot{x}(t) = Ax(t) + b(u(t) + \sigma(t)), \quad x(0) = x_0; \quad y(t) = c^T x(t) \tag{F.7}$$

where $x(t) \in \mathbb{R}^n$.

For feedback, a strictly proper and stable low-pass filter $C(s)$ with DC-gain of $0 \ dB$ is applied. Its bandwidth is one of the design parameters. The feedforward signal is filtered by $C_r(s)$ which is also strictly proper and stable with DC-gain of $0 \ dB$.

Let $H(s)$ be defined as:

$$H(s) := \frac{F(s)M(s)}{C(s)F(s) + (1 - C(s))M(s)} \tag{F.8}$$

Let $H_0(s)$ be defined as:

$$H_0(s) := \frac{F(s)}{C(s)F(s) + (1 - C(s))M(s)} \tag{F.9}$$

Let $H_1(s)$ be defined as:

$$H_1(s) := \frac{C(s)(F(s) - M(s))}{C(s)F(s) + (1 - C(s))M(s)} \tag{F.10}$$

Let $H_{1,r}(s)$ be defined as:

$$H_{1,r}(s) := \frac{C_r(s)(F(s) - M(s))}{C(s)F(s) + (1 - C(s))M(s)} \tag{F.11}$$

Let $H_2(s)$ be defined as:

$$H_2(s) := \frac{C(s)}{M(s)} H(s) \tag{F.12}$$

Let $H_3(s)$ be defined as:

$$H_3(s) := \frac{C(s)M(s)}{C(s)F(s) + (1 - C(s))M(s)} \tag{F.13}$$

Lemma **F.1.1.**

The transfer functions $H(s)$, $H_0(s)$, $H_1(s)$, $H_2(s)$, $H_3(s)$ are proper or strictly proper. The proof can be found in section F.5.1.

Lemma **F.1.2.**

The transfer functions $H(s)$, $H_0(s)$, $H_1(s)$, $H_2(s)$, $H_3(s)$ are stable. The proof can be found in section F.5.2.

To facilitate the stability proof for the reference system, the choice of $M(s)$ and $C(s)$ is subject to the $\mathcal{L}_1$-norm condition:

$$\|H(s)(1 - C(s))\|_{\mathcal{L}_1} L < 1 \tag{F.14}$$

where L is defined in Assumption F.1.4.

Remark **F.1.3.**

The $\mathcal{L}_1$-norm condition is necessary for the stability of the reference system later defined in section F.2 that assumes a perfect adaptive law. Hence, it is independent of the sampling time. The sampling time is a dependence in the deviation from the reference system as shown later.

Further define the 1_1-vector:

$$1_1 := [1, 0, ..., 0]^T \in \mathbb{R}^n \tag{F.15}$$

Lemma F.1.3.

Define $P = P^T > 0$ as the solution and $Q > 0$ as the parameter of the algebraic Lyapunov equation with the system matrix A. Then:

$$PA + A^T P = -Q$$

A simple choice of Q is I. See section F.5.3 for an explanation.

Define a matrix D, such that:

$$D \left(c^T \sqrt{P}^{-1} \right)^T = 0 \tag{F.16}$$

where the matrix $P = P^T > 0$, with $\sqrt{P}$ is such that $\sqrt{P}^T \sqrt{P} = P$.

Then, D is called the left nullspace or the cokernel of $\left(c^T \sqrt{P}^{-1} \right)^T$.

By using this definition the matrix Λ is defined as:

$$\Lambda := \begin{bmatrix} c^T \\ D\sqrt{P} \end{bmatrix} \tag{F.17}$$

By the definition of the nullspace it can be concluded that:

$$\Lambda \sqrt{P}^{-1} = \begin{bmatrix} c^T \sqrt{P}^{-1} \\ D \end{bmatrix} \tag{F.18}$$

has full rank, meaning that Λ is invertible.

Further define the following controller variables:

$$\Phi(t) := \int_0^t e^{\Lambda A \Lambda^{-1}(t-\tau)} \Lambda d\tau = \left(-\Lambda A \Lambda^{-1} \right)^{-1} \left(\mathbb{I} - e^{\Lambda A \Lambda^{-1} t} \right) \Lambda \tag{F.19}$$

where $\mathbb{I}$ is the unit-matrix in $\mathbb{R}^{n \times n}$.

Define the matrix $\Upsilon(t)$ as:

$$\Upsilon(t) := e^{\Lambda A \Lambda^{-1} t} \tag{F.20}$$

Define $\tilde{x}(t) := \hat{x}(t) - x(t) - \nu'(t)$, such that $\Lambda(\hat{x}(t) - x(t) - \nu'(t)) = [\hat{y}(t) - y(t) - \nu(t),\ \tilde{z}]^T$ and $\tilde{y}(t) := \hat{y}(t) - y(t) - \nu(t)$,

where the state transformation matrix Λ is defined such that:

$$\tilde{\xi}(t) = \Lambda \tilde{x}(t) = \begin{bmatrix} \tilde{y}(t) \\ \tilde{z}(t) \end{bmatrix} \tag{F.21}$$

Define an output predictor as:

$$\dot{\hat{x}}(t) = A\hat{x}(t) + bu(t) + \hat{\sigma}(t), \quad \hat{x}(0) = \hat{x}_0; \quad \hat{y}(t) = c^T \hat{x}(t) \tag{F.22}$$

The output predictor in Laplace domain is:

$$\hat{y}(s) = M(s)u(s) + M_{um}(s)\hat{\sigma}(s) \tag{F.23}$$

with $M(s) := c^T(s\mathbb{I} - A)^{-1}b$
and $M_{um}(s) := c^T(s\mathbb{I} - A)^{-1}$

Note that $M_{um}(s)$ is a row vector of transfer functions.

The adaptive law for updating the variable $\hat{\sigma}(t)$ is defined as:

$$\hat{\sigma}(t) = \hat{\sigma}(iT) := -\Phi(T)^{-1}\Upsilon(T)1_1\tilde{y}(iT) + \Phi^{-1}(T)1_1 h(iT), \quad i = 0, 1, 2, \ldots \tag{F.24}$$

where:

$$h(iT) = -\tilde{y}(iT) + h((i-1)T), \quad h(0) = 0,\ i = 1, 2, \ldots \tag{F.25}$$

meaning that the recursion starts with $h(0) = 0$ and continues with $h(1T) = -\tilde{y}(1T)$.

Note that:

$$\Upsilon(T)\Upsilon^{-1}(T)1_1 h(iT) = [h(iT), 0, \ldots, 0]^T \in \mathbb{R}^n \tag{F.26}$$

Due to the observer nature of this concept, $\hat{\sigma}(t)$ is vectorial and therefore not expressed via b in contrast to the plant description's scalar uncertainty.

With $\tilde{x}(t) = \hat{x}(t) - x(t) - \nu'(t)$, the error dynamics without measurement noise are defined as:

$$\dot{\tilde{x}}(t) = A\tilde{x}(t) - b\sigma(t) + \hat{\sigma}(t), \quad \tilde{x}(0) = \tilde{x}_0; \quad \tilde{y}(t) = c^T \tilde{x}(t) \tag{F.27}$$

The control law is defined as:

$$u(s) = C_r(s)r(s) - C(s)\frac{M_{um}(s)}{M(s)}\hat{\sigma}(s) \tag{F.28}$$

$M_{um}(s)$ defines the map from $\hat{\sigma}(s)$ to $\hat{y}(s)$, and $M^{-1}(s)$ the map from $\hat{y}(s)$ to $u(s)$. This is how the vector $\hat{\sigma}(s)$ is mapped to the input $u(s)$. The uncertainty is "visible" between $u(s)$ and $y(s)$, but $\hat{\sigma}(s)$ is applied to the states $\hat{x}(s)$.

The command $r(s)$ can be used either as digital signal as in a FBW-system, or as some direct mechanical input that is measured, or as baseline control signal. Both values enter the input equation in the same way. See Chapter 3.1 for additional comments.

Further, define $v(iT)$:

$$v(iT) = \begin{bmatrix} v_1(iT) \\ v_2(iT) \end{bmatrix} := \Upsilon^{-1}(T)1_1 h(iT) + \begin{bmatrix} 0 \\ \tilde{z}(iT) \end{bmatrix} \tag{F.29}$$

where $v_1(t) \in \mathbb{R}$ and $v_2(t) \in \mathbb{R}^{n-1}$.

If an exogenous input $u_o(t)$ is assumed to be present, e.g. the baseline control signal or the manual command from a pilot, there are no conditions on this input as persistent excitation. It is assumed however that it causes a stable response, i.e. there exist L_u and L_{u0} for $\|u_o\|_{\mathcal{L}_\infty} \leq \bar{u}_o$ such that:

$$\|y\|_{\mathcal{L}_\infty} \leq L_u \bar{u}_o + L_{u0} \tag{F.30}$$

Then:

$$\|d_o\|_{\mathcal{L}_\infty} \leq L\left(L_u \bar{u}_o + L_{u0}\right) + L_0 = \bar{d}_o \tag{F.31}$$

In broken-loop mode, plant and predictor are amended by initial conditions:

The plant equation is then:

$$y(s) = M(s)(u_o(s) + \sigma(s)) + y_{in}(s) \tag{F.32}$$

where $y_{in}(s)$ is the output of the following system:

$$\dot{x}_{in}(t) = A_p x_{in}(t), \quad x_{in}(0) = x_0; \quad y_{in}(t) = c_p^T x_{in}(t) \tag{F.33}$$

Here, A_p and c_p stem from the state space formulation of the plant $F(s)$.

Likewise, the output predictor has a nonzero initial condition:

$$\hat{y}(s) = M(s)u_o(s) + M_{um}(s)\hat{\sigma}(s) + \hat{y}_{in}(s) \tag{F.34}$$

where $\hat{y}_{in}(s)$ is the output of the following system:

$$\dot{\hat{x}}_{in}(t) = A\hat{x}_{in}(t), \ \hat{x}_{in}(0) = \hat{x}_0; \quad \hat{y}_{in}(t) = c^T \hat{x}_{in}(t) \tag{F.35}$$

Define $\rho_{\hat{x}_0}$ such that $\|\hat{x}_0\|_\infty \le \rho_{\hat{x}_0}$ and
define ρ_{x_0} such that $\|x_0\|_\infty \le \rho_{x_0}$.

Then $\rho_{\tilde{x}_0}$ is such that $\|\tilde{x}_0\|_\infty \le \rho_{\tilde{x}_0}$ and $\rho_{\tilde{y}_{in}}$ such that $\rho_{\tilde{y}_{in}} = \sup_{\rho_{\hat{x}_0},\rho_{x_0}} \|\tilde{y}_{in}\|_{\mathcal{L}_\infty}$

Further define the following:

$$\Delta := \|H_1(s)\|_{\mathcal{L}_1} \|r\|_{\mathcal{L}_\infty} + \|H_0(s)\|_{\mathcal{L}_1} (L\rho_r + L_0)$$
$$+ \left(\left\| \frac{H_1(s)}{M(s)} \right\|_{\mathcal{L}_1} + \|H_0(s)\|_{\mathcal{L}_1} L \frac{\|H_2(s)\|_{\mathcal{L}_1}}{1 - \|H(s)(1 - C(s))\|_{\mathcal{L}_1} L} \right) \bar{\gamma}_0 \tag{F.36}$$

with $\bar{\gamma}_0 > 0$ being an arbitrary constant, chosen freely.

In broken-loop mode, Δ is defined to be:

$$\Delta_o := \left(\left\| \frac{F(s) - M(s)}{M(s)} \right\|_{\mathcal{L}_1} + \left\| \frac{F(s)}{M(s)} \right\|_{\mathcal{L}_1} LL_u \right) \bar{u}_o + \left\| \frac{F(s)}{M(s)} \right\|_{\mathcal{L}_1} (LL_{u0} + L_0) \tag{F.37}$$

Further define:

$$\alpha := max \left(\lambda_{max}(\Lambda^{-T}P\Lambda^{-1}) \left(\frac{2 \left\| \Lambda^{-T}Pb \right\|_2 max(\Delta, \Delta_o)}{\lambda_{min} (\Lambda^{-T}Q\Lambda^{-1})} \right)^2, \ \lambda_{max}(K_2) \|\Lambda\|_\infty^2 n\rho_{\tilde{x}_0}^2 \right) \tag{F.38}$$

$$\varsigma := \left\| 1_1^T \Upsilon(T) \right\|_2 \sqrt{\frac{\alpha}{\lambda_{min} (\Lambda^{-T}P\Lambda^{-1})}} + \int_0^T \left| 1_1^T e^{\Lambda A \Lambda^{-1}(T-\tau)} \Lambda b \right| d\tau \, max(\Delta, \Delta_o) + \bar{\nu}_{rec} \tag{F.39}$$

A positive definite function $V(t)$ is defined as:

$$V(t) := w^T(t)\Lambda^{-T}P\Lambda^{-1}w(t) \tag{F.40}$$

with $w(t)$ being a vector of some states.

The bounds will be found as:

$$\rho_r := \frac{\|H(s)C_r(s)\|_{\mathcal{L}_1}\|r\|_{\mathcal{L}_\infty} + \|H(s)(1-C(s))\|_{\mathcal{L}_1} L_0}{1 - \|H(s)(1-C(s))\|_{\mathcal{L}_1} L} \tag{F.41}$$

$$
\begin{aligned}
\gamma_0 := &\max_{t\in[0,T]}\left(1_1^T e^{\Lambda A\Lambda^{-1}t}\right)\sqrt{\frac{\alpha}{\lambda_{min}\left(\Lambda^{-T}P\Lambda^{-1}\right)}} \\
&+ \max_{t\in[0,T]}\left(\int_0^t \left|1_1^T e^{\Lambda A\Lambda^{-1}(t-\tau)}\Lambda\Phi^{-1}(T)\Upsilon(T)1_1\right|d\tau\right)\varsigma' \\
&+ \max_{t\in[0,T]}\left(\int_0^t \left|1_1^T e^{\Lambda A\Lambda^{-1}(t-\tau)}\Lambda\Phi^{-1}(T)1_1\right|d\tau\right)\|h'(iT)\|_2 \\
&+ \max_{t\in[0,T]}\left(\int_0^t \left|1_1^T e^{\Lambda A\Lambda^{-1}(t-\tau)}\Lambda b\right|d\tau\right)\Delta + \bar{\nu}_{rec}
\end{aligned}
\tag{F.42}$$

$$
\begin{aligned}
\gamma_{0,o} := &\max_{t\in[0,T]}\left(1_1^T e^{\Lambda A\Lambda^{-1}t}\right)\sqrt{\frac{\alpha}{\lambda_{min}\left(\Lambda^{-T}P\Lambda^{-1}\right)}} \\
&+ \max_{t\in[0,T]}\left(\int_0^t \left|1_1^T e^{\Lambda A\Lambda^{-1}(t-\tau)}\Lambda\Phi^{-1}(T)\Upsilon(T)1_1\right|d\tau\right)\varsigma' \\
&+ \max_{t\in[0,T]}\left(\int_0^t \left|1_1^T e^{\Lambda A\Lambda^{-1}(t-\tau)}\Lambda\Phi^{-1}(T)1_1\right|d\tau\right)\|h'(iT)\|_2 \\
&+ \max_{t\in[0,T]}\left(\int_0^t \left|1_1^T e^{\Lambda A\Lambda^{-1}(t-\tau)}\Lambda b\right|d\tau\right)\Delta_o + \bar{\nu}_{rec}
\end{aligned}
\tag{F.43}$$

where $h'(t)$ and ς' are hypothetical noise-free $h(t)$ and ς.

$$\gamma_1 := \frac{\|H_2(s)\|_{\mathcal{L}_1}}{1 - \|H(s)(1-C(s))\|_{\mathcal{L}_1} L}\bar{\gamma}_0 \tag{F.44}$$

$$\gamma_2 := \|H_2(s)\|_{\mathcal{L}_1} L\gamma_1 + \left\|\frac{H_3(s)}{M(s)}\right\|_{\mathcal{L}_1}\bar{\gamma}_0 \tag{F.45}$$

$$\gamma_{2,o} := \left\|\frac{C(s)}{M(s)}\right\|_{\mathcal{L}_1}\bar{\gamma}_{0,o} + \left\|\frac{C(s)}{M(s)}\right\|_{\mathcal{L}_1}\rho_{\tilde{y}_{in}} \tag{F.46}$$

F.2 Reference System

A reference system is established, which formulates a closed-loop plant while assuming a perfect adaptive law, i.e. the uncertainty $\sigma(s)$ is assumed to be known due to the hypothetically known disturbance $d_{ref}(s)$ and dynamics $F(s)$. The purpose of introducing this reference system is to formulate stability independent of the adaptive law. After formulating stability conditions for the reference system, an upper bound for the deviation of the real system from the reference system is found that includes the adaptation quality in terms of the computational time step length T.

In contrast to the desired system however it includes the frequency limitations of control by the low-pass filter, i.e. a desired model that is supplemented by degradation of low-frequency control.

$$y_{ref}(s) = M(s)(u_{ref}(s) + \sigma_{ref}(s)) + y_{in}(s) \tag{F.47}$$

$$u_{ref}(s) = C_r(s)r(s) - C(s)\sigma_{ref}(s) \tag{F.48}$$

$$\sigma_{ref}(s) = \frac{(F(s) - M(s))u_{ref}(s) + F(s)d_{ref}(s)}{M(s)} \tag{F.49}$$

Lemma F.2.1.

This reference system is BIBO-stable if the $\mathcal{L}_1$-norm condition (F.14) is fulfilled, i.e.:

$$\|y_{ref,i}\|_{\mathcal{L}_\infty} \leq \rho_r$$

The proof can be found in section F.5.4.

F.3 Core Proof in Initial Broken-Loop Mode

This section addresses the output feedback controller initialization in the presence of sensor noise.

It is sufficient to show initialization effects in broken-loop mode – provided that the broken-loop mode is maintained for a sufficient time span. Switching to closed-loop mode affects the $\mathcal{L}_1$-controller only in the sense that new control signals are sent to the predictor but it does not differ from normal operation regarding internal states.

Bounds for the error dynamics regarding output and internal states are derived. With these bounds, a bound for the deviation from the reference system is found. As there is no direct control mechanism for the internal states as there is for the output, it is shown with a positive definite function (similar to the concept of Lyapunov functions) that the error in the internal states is bounded. The separation of the output from internal states is done by a state transformation matrix Λ. This matrix permits the segregation of the proof into two segments.

Some of the following considerations are based on [33] and adopted to this proof.

It is proved that $\|\tilde{y}\|_{\mathcal{L}_\infty}$, $\|y_{ref} - y\|_{\mathcal{L}_\infty}$, $\|u_{ref} - u\|_{\mathcal{L}_\infty}$ exist:

$$\|\tilde{y}\|_{\mathcal{L}_\infty} < \bar{\gamma}_{0,o} \tag{F.50}$$

$$\|y_{ref} - y\|_{\mathcal{L}_\infty} = 0 \tag{F.51}$$

$$\|u_{ref} - u\|_{\mathcal{L}_\infty} < \gamma_{2,o} \tag{F.52}$$

First, it is shown that the controller is stable in broken-loop mode. This refers to error dynamics and the boundedness of the control signal which is computed but discarded. Second, the portion of the control signal contributed by the initialization error is quantified and shown to be decaying.

The internal states of the predictor are initialized with some $\hat{x}(0)$ such that $c^T \hat{x}(0) = y(0)$, $\|\hat{x}_0\|_\infty \leq \rho_{\hat{x}_0}$, thus $\tilde{y}(0) = 0$ and in general $\tilde{x}(0) \neq 0$. Furthermore, the recursive term of the adaptive law is set $h(0) = 0$.

A proof by contradiction is not necessary as the upper bound for $\sigma(t)$ can be found straightforwardly. See also the comment about necessity of proof by contradiction on page 153.

The new error dynamics are obtained by subtracting (F.32) from (F.34):

$$\tilde{y}(s) = \hat{y}(s) - y(s) - \nu(s) = M_{um}(s)\hat{\sigma}(s) - M(s)\sigma(s) + \tilde{y}_{in}(s) + \nu_{rec}(s) \tag{F.53}$$

This – without sensor noise – can be written in state space formulation as:

$$\dot{\tilde{x}}(t) = A\tilde{x}(t) - b\sigma(t) + \hat{\sigma}(t), \quad \tilde{x}(0) = \tilde{x}_0; \quad \tilde{y}(t) = c^T\tilde{x}(t) \tag{F.54}$$

Transforming this as described in (F.21) delivers:

$$\dot{\tilde{\xi}}(t) = \Lambda A \Lambda^{-1}\tilde{\xi}(t) - \Lambda b \sigma(t) + \Lambda \hat{\sigma}(t), \quad \tilde{\xi}(0) = \tilde{\xi}_0; \quad \tilde{y}(t) = \tilde{\xi}_1(t) \tag{F.55}$$

Note that $\tilde{\xi}(0) = \Lambda \tilde{x}_0$ and $\tilde{\xi}_1(0) = 0$.

In this equation, $\tilde{\xi}_1(t)$ denotes the first element of the vector $\tilde{\xi}(t)$. Equation (F.55) has the following solution with sensor noise and with $t_0 = iT$ for $t < T$, i.e. $\hat{\sigma}(iT)$ in the equation is seen as a constant.:

$$\tilde{\xi}(iT + t) = e^{\Lambda A \Lambda^{-1}t}\tilde{\xi}(iT) + \int_{iT}^{iT+t} e^{\Lambda A \Lambda^{-1}(iT+t-\tau)}\Lambda \hat{\sigma}(iT)d\tau \tag{F.56}$$
$$- \int_{iT}^{iT+t} e^{\Lambda A \Lambda^{-1}(iT+t-\tau)}\Lambda b \sigma(\tau)d\tau + \Lambda \nu_{\tilde{x}}(iT)$$

This can be rewritten to:

$$\tilde{\xi}(iT + t) = e^{\Lambda A \Lambda^{-1}t}\tilde{\xi}(iT) + \int_{0}^{t} e^{\Lambda A \Lambda^{-1}(t-\tau)}\Lambda \hat{\sigma}(iT)d\tau \tag{F.57}$$
$$- \int_{0}^{t} e^{\Lambda A \Lambda^{-1}(t-\tau)}\Lambda b \sigma(iT + \tau)d\tau + \Lambda \nu_{\tilde{x}}(iT)$$

Now split Equation (F.57) into two components for separating $\tilde{y}(iT)$ and $\hat{\sigma}(iT)$ from the rest, then with (F.21):

$$\chi_{raw}(iT + t) = e^{\Lambda A \Lambda^{-1}t}\begin{bmatrix} \tilde{y}(iT) \\ 0_{(n-1)} \end{bmatrix} + \int_{0}^{t} e^{\Lambda A \Lambda^{-1}(t-\tau)}\Lambda \hat{\sigma}(iT)d\tau \tag{F.58}$$

$$\zeta_{raw}(iT + t) = e^{\Lambda A \Lambda^{-1}t}\begin{bmatrix} 0 \\ \tilde{z}(iT) \end{bmatrix} - \int_{0}^{t} e^{\Lambda A \Lambda^{-1}(t-\tau)}\Lambda b \sigma(iT + \tau)d\tau \tag{F.59}$$

The components are such that:

$$\tilde{\xi}(iT + t) = \chi_{raw}(iT + t) + \zeta_{raw}(iT + t) + \Lambda \nu_{\tilde{x}}(iT) \tag{F.60}$$

> The upcoming proof of boundedness for $\tilde{y}(t)$ and $\tilde{z}(t)$ takes place in three steps.
>
> **First,** boundedness is proved for $t = 0$.
>
> **Second,** if from boundedness at one point of time $t = i_0 T$ follows boundedness at $t = (i_0 + 1)T$, then for all $t = iT$ boundedness prevails.
>
> **Third,** all remaining times are evaluated by $\tilde{y}(iT + t)$.

Remark F.3.1.

An equation containing some dependency on iT describes a set of equations, one for every i. In the following, one not further specified $i_0 T$ is taken which is *one* choice of the set given by iT, according to $i = 1, 2, \ldots$.

Lemma F.3.1.

Assume a vector $w \in \mathbb{R}^n$ where $w = [w_1, w_2]^T$ with $w_1 \in \mathbb{R}$ and $w_2 \in \mathbb{R}^{n-1}$, then there exists a constant $k_1 \in \mathbb{R}$ and a matrix $K_2 \in \mathbb{R}^{(n-1)\times(n-1)}$ so that:

$$w^T \Lambda^{-T} P \Lambda^{-1} w = k_1 w_1^2 + w_2^T K_2 w_2$$

The proof of this Lemma can be found in section F.5.5.

First, the initial condition of $h(0) = 0$ ensures decoupling of $\tilde{y}(0)$ from $\tilde{z}(0)$, i.e. $v(0)$ collapses to $[\tilde{y}(0), \tilde{z}(0)]^T$. Note that $\tilde{y}(0) = 0 \leq \varsigma(t)$.

Lemma F.3.2.

Let $\lambda_{min}(A)$ and $\lambda_{max}(A)$ be the minimum and the maximum eigenvalue of the positive definite $(A > 0)$ square matrix A, respectively. Then:

$$\lambda_{min}(A) \|x\|_2^2 \leq x^T A x \leq \lambda_{max}(A) \|x\|_2^2$$

The proof is shown in section F.5.6.

Then, with $\tilde{\xi}(0) = [0, \tilde{z}(0)]^T$, Lemma F.3.1 can be used to gain $\tilde{z}^T(0)K_2\tilde{z}(0) \leq \alpha$. This follows from

$$\tilde{\xi}^T(0)K_2\tilde{\xi}(0) \leq \lambda_{max}(K_2)\left\|\tilde{\xi}(0)\right\|_2^2 \leq \lambda_{max}(K_2)\|\Lambda\tilde{x}_0\|_2^2 \leq \lambda_{max}(K_2)\|\Lambda\|_\infty^2\, n\rho_{\tilde{x}_0}^2 \leq \alpha \tag{F.61}$$

and the fact that $\tilde{\xi}(0) = \tilde{z}(0)$.

Second, $t = i_0 T$:

The following describes the examination whether from boundedness at $t = i_0 T$ follows boundedness at $t = (i_0 + 1)T$.

Hence it is assumed that:

$$\tilde{y}(i_0 T) \leq \varsigma, \quad V(i_0 T) \leq \alpha \tag{F.62}$$

where ς and α were defined in Equations (F.39) and (F.38).

Equations (F.58) and (F.59) for $t = (i_0 + 1)T$ are:

$$\chi_{raw}((i_0 + 1)T) = e^{\Lambda A \Lambda^{-1}T}\begin{bmatrix}\tilde{y}(i_0 T) \\ 0_{(n-1)}\end{bmatrix} + \int_0^T e^{\Lambda A \Lambda^{-1}(T-\tau)}\Lambda\hat{\sigma}(i_0 T)d\tau \tag{F.63}$$

$$\zeta_{raw}((i_0 + 1)T) = e^{\Lambda A \Lambda^{-1}T}\begin{bmatrix}0 \\ \tilde{z}(i_0 T)\end{bmatrix} - \int_0^T e^{\Lambda A \Lambda^{-1}(T-\tau)}\Lambda b\sigma(i_0 T + \tau)d\tau \tag{F.64}$$

Equation (F.63) can be written as:

$$\chi_{raw}((i_0 + 1)T) = \Upsilon(T)\begin{bmatrix}\tilde{y}(i_0 T) \\ 0_{(n-1)}\end{bmatrix} + \Phi(T)\hat{\sigma}(i_0 T) \tag{F.65}$$

Merging the adaptive law (F.24) into (F.65) the following is true:

First:

$$\chi_{raw}((i_0 + 1)T) = 0 + \Phi(T)\Phi^{-1}(T)1_1 h(i_0 T) \tag{F.66}$$

Second:

$$\zeta_{raw}((i_0 + 1)T) = \Upsilon(T) \begin{bmatrix} 0 \\ \tilde{z}(i_0 T) \end{bmatrix} - \int_0^T e^{\Lambda A \Lambda^{-1}(T-\tau)} \Lambda b \sigma(i_0 T + \tau) d\tau \tag{F.67}$$

The term $\Phi(T)\Phi^{-1}(T)1_1 h(i_0 T) = 1_1 h(i_0 T)$ from the equation for $\chi_{raw}((i_0 + 1)T)$ (where it originally belongs to because of $\hat{\sigma}(i_0 T)$) is now pulled into the equation of $\zeta_{raw}((i_0 + 1)T)$ in the form of $\Upsilon(T)\Upsilon^{-1}(T)1_1 h(i_0 T)$ such that $\zeta((i_0 + 1)T) = \zeta_{raw}(iT) + \Upsilon(T)\Upsilon^{-1}(T)1_1 h(i_0 T)$. This is allowable because of Equation (F.60), in which a term from one summand is shifted to another summand – the sum is the same.

Hence:

$$\chi((i_0 + 1)T) = \chi_{raw}((i_0 + 1)T) - 1_1 h(i_0 T) = 0 \tag{F.68}$$

as well as:

$$\zeta((i_0 + 1)T) = \zeta_{raw}(i_0 T + t) + \Upsilon(T)\Upsilon^{-1}(T)\Phi(T)\Phi^{-1}(T)1_1 h(i_0 T) \tag{F.69}$$

Equation (F.69) is reformulated with the help of (F.64) to:

$$\zeta((i_0 + 1)T) = \Upsilon(T) \begin{bmatrix} 0 \\ \tilde{z}(i_0 T) \end{bmatrix} - \int_0^T e^{\Lambda A \Lambda^{-1}(T-\tau)} \Lambda b \sigma(i_0 T + \tau) d\tau + \Upsilon(T)\Upsilon^{-1}(T)1_1 h(i_0 T) \tag{F.70}$$

In an adaptive law not featuring the additional term with $h(iT)$, Equation (F.64) states the remaining term. This remaining term arises in the time span where the old cumulated uncertainties are canceled while new uncertainties are cumulating again. See Chapter 2.5.2 for a more precise explanation. With the recursive definition of $h(iT)$ based on $\tilde{y}(iT)$, it can be seen that these remaining terms are taken into account via $h(iT)$. Note that $h(iT)$ is scalar:

$$h(iT) = -1_1^T e^{\Lambda A \Lambda^{-1} T} \begin{bmatrix} 0 \\ \tilde{z}'((i-1)T) \end{bmatrix} + 1_1^T \int_0^T e^{\Lambda A \Lambda^{-1}(T-\tau)} \Lambda b \sigma((i-1)T + \tau) d\tau + \nu_h(iT) \tag{F.71}$$

where $\tilde{z}'(t)$ is a hypothetical noise-free $\tilde{z}(t)$ to ensure uniqueness of noise in $\nu_h(iT)$.

Equation (F.70) can be rewritten to:

$$\zeta((i_0 + 1)T) = \Upsilon(T) \begin{bmatrix} v_1(i_0T) \\ v_2(i_0T) \end{bmatrix} - \int_0^T e^{\Lambda A \Lambda^{-1}(T-\tau)} \Lambda b \sigma(i_0T + \tau) d\tau \qquad (F.72)$$

where $v_1(t)$ and $v_2(t)$ are defined in Equation (F.29).

Equation (F.72) is the solution of the following differential equation, where $t \in [i_0T, (i_0 + 1)T)$:

$$\dot{\zeta}(t) = \Lambda A \Lambda^{-1} \zeta(t) - \Lambda b \sigma(t) \qquad (F.73)$$

with the initial condition of:

$$\zeta(i_0T) = \begin{bmatrix} v_1(i_0T) \\ v_2(i_0T) \end{bmatrix} = v(i_0T) \qquad (F.74)$$

The positive definite function $V(t)$ was defined according to (F.40) as:

$$V(t) = \zeta^T(t) \Lambda^{-T} P \Lambda^{-1} \zeta(t) \qquad (F.75)$$

With Λ being invertible (see Equation (F.18)) and $P > 0$ (see section F.5.3), it follows that $V > 0$ with $\Lambda^{-T} P \Lambda^{-1} > 0$.

The relocation of $h(i_0T)$ stated in Equations (F.66) and (F.68) in comparison to [13] is possible as the additional term $h(i_0T)$ is piece-wise constant, and therefore can be merged into the initial condition for a positive definite function $V(t)$ on the defined time length. This is true as the function $V(t)$ is used to examine the behavior in a continuous interval $t \in [i_0T, (i_0 + 1)T)$ over which $h(i_0T)$ is constant.

Applying Lemma F.3.1, for the positive definite function $V(t)$ at $t = i_0T$ the following is true:

$$V(i_0T) = k_1 \zeta_1^2(i_0T) + \zeta_2^T(i_0T) K_2 \zeta_2(i_0T) \qquad (F.76)$$

With the assumption in (F.62) that $V(i_0T) \leq \alpha$ (needed to see whether this implies boundedness for $t = (i_0 + 1)T$ as well):

$$V(i_0T) \leq \alpha \qquad (F.77)$$

For examining boundedness of $V(t)$ for $t \geq i_0 T$, the derivative is examined:

$$\dot{V}(t) = \zeta^T(t)\Lambda^{-T}P\Lambda^{-1}\dot{\zeta}(t) + \dot{\zeta}^T(t)\Lambda^{-T}P\Lambda^{-1}\zeta(t) \tag{F.78}$$

With Equation (F.73):

$$\dot{V}(t) = \zeta^T(t)\Lambda^{-T}P\Lambda^{-1}(\Lambda A\Lambda^{-1}\zeta(t) - \Lambda b\sigma) + (\Lambda A\Lambda^{-1}\zeta(t) - \Lambda b\sigma)^T\Lambda^{-T}P\Lambda^{-1}\zeta(t) \tag{F.79}$$

$$\dot{V}(t) = \zeta^T(t)\Lambda^{-T}(PA + A^TP)\Lambda^{-1}\zeta(t) - 2\zeta^T(t)\Lambda^{-T}Pb\sigma(t) \tag{F.80}$$

where Lemma F.1.3 can be applied:

$$PA + A^TP = -Q \tag{F.81}$$

From Assumption F.1.4 and from (F.30) it follows (F.31):

$$\|d_o\|_{\mathcal{L}_\infty} \leq L\left(L_u\bar{u}_o + L_{u0}\right) + L_0 = \bar{d}_o \tag{F.82}$$

From the definition of $\sigma(s)$ in (F.6) one has:

$$\|\sigma\|_{\mathcal{L}_\infty} \leq \left\|\frac{F(s) - M(s)}{M(s)}\right\|_{\mathcal{L}_1}\bar{u}_o + \left\|\frac{F(s)}{M(s)}\right\|_{\mathcal{L}_1}\bar{d}_o = \Delta_o \tag{F.83}$$

And with Equation (F.82):

$$\|\sigma\|_{\mathcal{L}_\infty} \leq \left(\left\|\frac{F(s) - M(s)}{M(s)}\right\|_{\mathcal{L}_1} + \left\|\frac{F(s)}{M(s)}\right\|_{\mathcal{L}_1}LL_u\right)\bar{u}_o + \left\|\frac{F(s)}{M(s)}\right\|_{\mathcal{L}_1}(LL_{u0} + L_0) = \Delta_o \tag{F.84}$$

Lemma F.3.2 delivers with inverted signs for a general matrix A and a vector x:

$$-\lambda_{min}(A)\|x\|_2^2 \geq -x^TAx \geq -\lambda_{max}(A)\|x\|_2^2 \tag{F.85}$$

Then:

$$\dot{V}(t) \leq -\lambda_{min}\left(\Lambda^{-T}Q\Lambda^{-1}\right)\|\zeta(t)\|_2^2 + 2\|\zeta(t)\|\left\|\Lambda^{-T}Pb\right\|_2\Delta_o \tag{F.86}$$

With $\Lambda^{-T}Q\Lambda^{-1}$ being positive definite (see explanation from Equation (F.75) where both, P and Q are positive definite, and the fact that a positive definite matrix always has only positive eigenvalues), the first term in Equation (F.86) can only be negative (including the $(-)$), whereas the second term is always positive.

Hence, the inequality $\dot{V}(t) < 0$ is fulfilled for:

$$\|\zeta\|_2 > \|\zeta_0\|_2 = \frac{2\left\|\Lambda^{-T}Pb\right\|_2 \Delta_o}{\lambda_{min}\left(\Lambda^{-T}Q\Lambda^{-1}\right)} \tag{F.87}$$

Using this result in (F.75) provides with the help of Lemma F.3.2:

$$V(\zeta_0) = \zeta_0(t)^T \Lambda^{-T} P \Lambda^{-1} \zeta_0(t) \leq \lambda_{max}(\Lambda^{-T}P\Lambda^{-1}) \left(\frac{2\left\|\Lambda^{-T}Pb\right\|_2 \Delta_o}{\lambda_{min}\left(\Lambda^{-T}Q\Lambda^{-1}\right)}\right)^2 \leq \alpha \tag{F.88}$$

where α is defined in (F.38).

This and $V(i_0 T) \leq \alpha$ show directly that:

$$V(t) \leq \alpha \quad \forall t \in [i_0 T, (i_0+1)T), \quad (i_0+1)T \leq \imath \tag{F.89}$$

With $\chi(i_0 T) = 0$ from Equation (F.68) and the fact that $\tilde{\xi}(t) = \zeta(t) + \chi(t)$ it follows that not only:

$$\zeta^T((i_0+1)T)\Lambda^{-T}P\Lambda^{-1}\zeta((i+1)T) \leq \alpha \tag{F.90}$$

but also:

$$\tilde{\xi}^T((i_0+1)T)\Lambda^{-T}P\Lambda^{-1}\tilde{\xi}((i+1)T) \leq \alpha \tag{F.91}$$

Using section F.5.5 with $\tilde{\xi}$ as the therein stated vector w, i.e.

$$\tilde{\xi}(i_0 T) = \zeta(i_0 T) = v(i_0 T) \tag{F.92}$$

where the last equal sign is due to Equation (F.74):

$$\tilde{\xi}((i_0+1)T)^T \Lambda^{-T}P\Lambda^{-1}\tilde{\xi}((i_0+1)T) = k_1 v_1^2((i_0+1)T) + v_2^T((i_0+1)T)K_2 v_2((i_0+1)T) \leq \alpha \tag{F.93}$$

which means the upper bound from Equation (F.77) holds for $t = (i_0 + 1)T \leq \iota$, too:

$$V((i_0 + 1)T) \leq \alpha \tag{F.94}$$

At this point it is proved that the second assumption of Equation (F.62), namely $V(i_0 T) \leq \alpha$ implies $V((i_0 + 1)T) \leq \alpha$, too. It is still to be proved that the same applies to the first assumption, namely from $\tilde{y}(i_0 T) \leq \varsigma$ implies $\tilde{y}((i_0 + 1)T) \leq \varsigma$. This is next:

From Equation (F.77) and with Lemma F.3.1 it follows that:

$$v^T(i_0 T)\Lambda^{-T}P\Lambda^{-1}v(i_0 T) \leq \alpha \tag{F.95}$$

This can be used with Lemma F.3.2 to obtain:

$$\lambda_{min}\left(\Lambda^{-T}P\Lambda^{-1}\right)\|v(i_0 T)\|_2^2 \leq v^T(i_0 T)\Lambda^{-T}P\Lambda^{-1}v(i_0 T) \tag{F.96}$$

Combining the last two equations delivers:

$$\lambda_{min}\left(\Lambda^{-T}P\Lambda^{-1}\right)\|v(i_0 T)\|_2^2 \leq \alpha \tag{F.97}$$

Which directly leads to:

$$\|v(i_0 T)\|_2 \leq \sqrt{\frac{\alpha}{\lambda_{min}\left(\Lambda^{-T}P\Lambda^{-1}\right)}} \tag{F.98}$$

Where:

$$\|v(i_0 T)\|_2 = \left\|\begin{bmatrix} v_1(i_0 T) \\ v_2(i_0 T) \end{bmatrix}\right\|_2 \tag{F.99}$$

meaning that $\begin{bmatrix} v_1(i_0 T), & v_2(i_0 T) \end{bmatrix}^T$ is bounded.

Next, as $\chi((i_0 + 1)T) = 0$:

$$\tilde{y}((i_0 + 1)T) = 1_1^T \zeta((i_0 + 1)T) + \nu_{rec}((i_0 + 1)T) \tag{F.100}$$

with the definition in (F.72):

$$\tilde{y}((i_0+1)T) = 1_1^T e^{\Lambda A \Lambda^{-1}T} \begin{bmatrix} v_1(i_0 T) \\ v_2(i_0 T) \end{bmatrix}' - 1_1^T \int_0^T e^{\Lambda A \Lambda^{-1}(T-\tau)} \Lambda b \sigma(i_0 T + \tau) d\tau + \nu_{rec}((i_0+1)T)$$

$$(F.101)$$

where $v'(t)$ is a hypothetical noise-free $v(t)$ to ensure uniqueness of noise in $\nu_{rec}(iT)$. Note that this equation states the closed loop behavior, i.e. with $\hat{\sigma}(t)$ being applied.

An upper bound can be derived with Equation (F.97) as:

$$|\tilde{y}((i_0+1)T)| \le \left\| 1_1^T \Upsilon(T) \right\|_2 \sqrt{\frac{\alpha}{\lambda_{min}\left(\Lambda^{-T} P \Lambda^{-1}\right)}}$$

$$+ \int_0^T \left| 1_1^T e^{\Lambda A \Lambda^{-1}(T-\tau)} \Lambda b \right| \left| \sigma(i_0 T + \tau) \right| d\tau + \bar{\nu}_{rec}$$

$$\le \varsigma$$

$$(F.102)$$

where according to (F.84) $|\sigma(i_0 T + \tau)| \le \Delta_o$. The boundedness of $v(t)$ is shown in (F.98), ς is defined in (F.39).

This means that from $\tilde{y}(i_0 T) \le \varsigma$ follows that $\tilde{y}((i+1)T) \le \varsigma$ as well, where $(i_0+1)T \le \wr$ and therefore $\forall i T \le \wr$.

With (F.99) and from the fact that $k_1 > 0$ as well as $K_2 > 0$ ('positive definite') it follows that both, $v_1(i_0 T)$ and $v_2(i_0 T)$ are bounded. In the definition of $v(t)$ in Equation (F.29) there is no influence of $\tilde{z}(i_0 T)$ to $v_1(i_0 T)$. As $v_1(i_0 T)$ is bounded and $h(i_0 T)$ is scalar, it follows that $h(i_0 T)$ is bounded as well. As the left side and the first summand of the equation are bounded, the second summand of the equation, namely $\tilde{z}(i_0 T)$ must be bounded, too. This argumentation can be repeated for $t = (i_0+1)T$ as well by the validity of boundedness of $V(i_0 T)$ and $y(i_0 T)$ $\forall i_0 \in i$.

Third, is the proof of $\tilde{y}(iT + t) \le \gamma_0$, where γ_0 is a constant and $0 \le t \le T$, $iT + t \le \wr$. The beginning t=0 is known, the assumption of bounds in (F.62) for $t = i_0 T$ with its conclusion that the bounds also hold for $t = (i_0+1)T$, and the addition of the times $t \in [iT, (i+1)T]$, $(i+1)T \le \wr$ provide a bound for all times $t \le \wr$. It is shown that a time $t = \wr$ where $\tilde{y}(\wr) = \bar{\gamma}_0$ is not reached.

Making use of Equation (F.57), one has:

$$\tilde{y}(iT + t) = 1_1^T e^{\Lambda A \Lambda^{-1}t} \tilde{\xi}(iT) + 1_1^T \int_0^t e^{\Lambda A \Lambda^{-1}(t-\tau)} \Lambda \hat{\sigma}(iT) d\tau$$

$$- 1_1^T \int_0^t e^{\Lambda A \Lambda^{-1}(t-\tau)} \Lambda b \sigma(iT + \tau) d\tau + \nu_{rec}(iT)$$

$$(F.103)$$

Note that for the integral $\hat{\sigma}(iT)$ is a constant as $0 \le t \le T$.

From Equation (F.92) it follows that $\tilde{\xi}(iT) = v(iT)$. Further, by using $\hat{\sigma}(iT)$ from the definition in (F.24) one has for $t \in [0, T)$:

$$\tilde{y}(iT + t) = 1_1^T \Upsilon(t) v'(iT) + 1_1^T \Phi(t)(-\Phi^{-1}(T)) \Upsilon(T) 1_1 \tilde{y}'(iT) + \tag{F.104}$$

$$1_1^T \Phi(t)\Phi^{-1}(T) 1_1 h'(iT) - 1_1^T \int_0^t e^{\Lambda A \Lambda^{-1}(t-\tau)} \Lambda b\sigma(iT + \tau) d\tau + \nu_{rec}(iT)$$

where $h'(t)$, $v'(t)$, $\tilde{y}'(t)$ and ς' are hypothetical noise-free $h(t)$, $v(t)$, $\tilde{y}(t)$ and ς to ensure uniqueness of noise in $\nu_{rec}(iT)$.

Now, $\tilde{y}(iT + t)$ for $t \in [0, T]$ is bounded by some constant γ_0, i.e. $\tilde{y}(iT + t) \le \gamma_0$:

$$
\begin{aligned}
\|\tilde{y}(iT + t)\|_\infty &\le \max_{t \in [0,T]} \left(1_1^T e^{\Lambda A \Lambda^{-1} t} \right) \sqrt{\frac{\alpha}{\lambda_{min}\left(\Lambda^{-T} P \Lambda^{-1} \right)}} \\
&+ \max_{t \in [0,T]} \left(\int_0^t \left| 1_1^T e^{\Lambda A \Lambda^{-1}(t-\tau)} \Lambda \Phi^{-1}(T) \Upsilon(T) 1_1 \right| d\tau \right) \varsigma' \\
&+ \max_{t \in [0,T]} \left(\int_0^t \left| 1_1^T e^{\Lambda A \Lambda^{-1}(t-\tau)} \Lambda \Phi^{-1}(T) 1_1 \right| d\tau \right) \|h'(iT)\|_2 \\
&+ \max_{t \in [0,T]} \left(\int_0^t \left| 1_1^T e^{\Lambda A \Lambda^{-1}(t-\tau)} \Lambda b \right| d\tau \right) \Delta_o + \bar{\nu}_{rec} \\
&= \gamma_{0,o}
\end{aligned}
\tag{F.105}
$$

It is shown on page 146 that $\|h(iT)\|_2$ exits.

Lemma F.3.3.

$$\lim_{T \to 0} \gamma_{0,o} = \nu_{rec}(t)$$

The proof is shown in section F.5.8.

Remark F.3.2.

In Equation (F.104) some terms show t, others T. To overcome assorted dependencies, the maximum in the span $t \in [0, T]$ is taken.

This proves stability of the broken-loop controller operation.

Next, the reference system is taken into account which helps quantifying an upper bound for transient errors in the control signal.

From (F.53) it follows that:

$$M_{um}(s)\hat{\sigma}(s) = \tilde{y}(s) + M(s)\sigma(s) - \tilde{y}_{in}(s) \qquad (\text{F.106})$$

With the controller in broken-loop mode, the reference system is identical to the plant dynamics. It follows that $y_{ref}(t) \equiv y(t)$ and $\sigma(t) \equiv \sigma_{ref}(t)$. Since the discarded control signal of the $\mathcal{L}_1$-controller is $u_{\mathcal{L}_1}(s) = C_r(s)r(s) - \frac{C(s)M_{um}(s)}{M(s)}\hat{\sigma}(s)$ and the reference system control signal is $u_{ref}(t) = C_r(s)r(s) - C(s)\sigma_{ref}(s)$ one has with Equation (F.106):

$$u_{ref}(s) - u_{\mathcal{L}_1}(s) = \frac{C(s)}{M(s)}\tilde{y}(s) - \frac{C(s)}{M(s)}\tilde{y}_{in}(s) \qquad (\text{F.107})$$

Hence:

$$\begin{aligned}
\|u_{ref} - u_{\mathcal{L}_1}\|_{\mathcal{L}_\infty} &\leq \left\|\frac{C(s)}{M(s)}\right\|_{\mathcal{L}_1} \|\tilde{y}\|_{\mathcal{L}_\infty} + \left\|\frac{C(s)}{M(s)}\right\|_{\mathcal{L}_1} \|\tilde{y}_{in}\|_{\mathcal{L}_\infty} \\
&< \left\|\frac{C(s)}{M(s)}\right\|_{\mathcal{L}_1} \bar{\gamma}_{0,o} + \left\|\frac{C(s)}{M(s)}\right\|_{\mathcal{L}_1} \rho_{\tilde{y}_{in}} \\
&= \gamma_{2,o}
\end{aligned} \qquad (\text{F.108})$$

The second term of Equation (F.108) offers indications for designing the duration of the broken-loop mode. With $\frac{C(s)}{M(s)}$ being proper and stable, the focus lies on the time constants of the exponentially fast decaying $\tilde{y}_{in}(s)$. As this time constant is determined by A, one or twice the rise time of the desired dynamics is proposed. For desired dynamics with $1\,Hz$ bandwidth and with the first order equivalent of the desired dynamics, the broken-loop mode would last for about $1...2\,s$, provided that the bandwidth of $C(s)$ is sufficiently fast (which it is expected to, as its bandwidth should exceed the one of desired dynamics) and the inverse of $M(s)$ is sufficiently fast, too.

F.4 Core Proof in Standard Closed-Loop Mode

In this section, the loop is closed and the adaptive input $u_{\mathcal{L}_1}(t)$ is applied, assuming that the term in the input signal due to the transient error has decayed sufficiently. The new $t = 0$ is now the first time step in closed-loop mode.

It is proved here that $\|\tilde{y}\|_{\mathcal{L}_\infty}$, $\|y_{ref} - y\|_{\mathcal{L}_\infty}$, $\|u_{ref} - u\|_{\mathcal{L}_\infty}$ exist:

$$\|\tilde{y}\|_{\mathcal{L}_\infty} < \bar{\gamma}_0 \qquad (\text{F.109})$$

$$\|y_{ref} - y\|_{\mathcal{L}_\infty} < \gamma_1 \tag{F.110}$$

$$\|u_{ref} - u\|_{\mathcal{L}_\infty} < \gamma_2 \tag{F.111}$$

Transforming Equation (F.27) for error dynamics as described in (F.21) delivers:

$$\dot{\tilde{\xi}}(t) = \Lambda A \Lambda^{-1} \tilde{\xi}(t) - \Lambda b \sigma(t) + \Lambda \hat{\sigma}(t), \quad \tilde{\xi}(0) = \tilde{\xi}_0; \quad \tilde{y}(t) = \tilde{\xi}_1(t) \tag{F.112}$$

In this equation, $\tilde{\xi}_1(t)$ denotes the first element of the vector $\tilde{\xi}(t)$. It has the following solution with $t_0 = iT$ for $t \le T$, i.e. $\hat{\sigma}(iT)$ in the equation is seen as a constant:

$$\tilde{\xi}(iT + t) = e^{\Lambda A \Lambda^{-1} t} \tilde{\xi}(iT) + \int_{iT}^{iT+t} e^{\Lambda A \Lambda^{-1}(iT+t-\tau)} \Lambda \hat{\sigma}(iT) d\tau \tag{F.113}$$
$$- \int_{iT}^{iT+t} e^{\Lambda A \Lambda^{-1}(iT+t-\tau)} \Lambda b \sigma(\tau) d\tau + \Lambda \nu_{\tilde{x}}(iT)$$

This can be rewritten to:

$$\tilde{\xi}(iT + t) = e^{\Lambda A \Lambda^{-1} t} \tilde{\xi}(iT) + \int_{0}^{t} e^{\Lambda A \Lambda^{-1}(t-\tau)} \Lambda \hat{\sigma}(iT) d\tau \tag{F.114}$$
$$- \int_{0}^{t} e^{\Lambda A \Lambda^{-1}(t-\tau)} \Lambda b \sigma(iT + \tau) d\tau + \Lambda \nu_{\tilde{x}}(iT)$$

Now split Equation (F.114) into two components for separating $\tilde{y}(iT)$ and $\hat{\sigma}(iT)$ from the rest, then with (F.21):

$$\chi_{raw}(iT + t) = e^{\Lambda A \Lambda^{-1} t} \begin{bmatrix} \tilde{y}(iT) \\ 0_{(n-1)} \end{bmatrix} + \int_{0}^{t} e^{\Lambda A \Lambda^{-1}(t-\tau)} \Lambda \hat{\sigma}(iT) d\tau \tag{F.115}$$

$$\zeta_{raw}(iT + t) = e^{\Lambda A \Lambda^{-1} t} \begin{bmatrix} 0 \\ \tilde{z}(iT) \end{bmatrix} - \int_{0}^{t} e^{\Lambda A \Lambda^{-1}(t-\tau)} \Lambda b \sigma(iT + \tau) d\tau \tag{F.116}$$

The components are such that:

$$\tilde{\xi}(iT + t) = \chi_{raw}(iT + t) + \zeta_{raw}(iT + t) + \Lambda \nu_{\tilde{x}}(iT) \tag{F.117}$$

> The upcoming proof of boundedness for $\tilde{y}(t)$ and $\tilde{z}(t)$ takes place in three steps.
>
> **First,** boundedness is proved for $t = 0$.
>
> **Second,** if from boundedness at one point of time $t = i_0 T$ follows boundedness at $t = (i_0 + 1)T$, then for all $t = iT$ boundedness prevails.
>
> **Third,** all remaining times are evaluated by $\tilde{y}(iT + t)$.

First, from the previous section it follows that $\tilde{\xi}(0) = \tilde{\xi}_0$ is bounded. Then, with (F.21) and Assumption F.1.3, boundedness of $\tilde{y}(0)$ and $\tilde{z}(0)$ is given.

Second, $t = i_0 T$:

The following describes the examination whether from boundedness at $t = i_0 T$ follows boundedness at $t = (i_0 + 1)T$.
Hence it is assumed that:

$$\tilde{y}(i_0 T) \leq \varsigma, \quad V(i_0 T) \leq \alpha \tag{F.118}$$

where ς and α were defined in Equations (F.39) and (F.38).

Equations (F.115) and (F.116) for $t = (i_0 + 1)T$ are:

$$\chi_{raw}((i_0 + 1)T) = e^{\Lambda A \Lambda^{-1} T} \begin{bmatrix} \tilde{y}(i_0 T) \\ 0_{(n-1)} \end{bmatrix} + \int_0^T e^{\Lambda A \Lambda^{-1}(T-\tau)} \Lambda \hat{\sigma}(i_0 T) d\tau \tag{F.119}$$

$$\zeta_{raw}((i_0 + 1)T) = e^{\Lambda A \Lambda^{-1} T} \begin{bmatrix} 0 \\ \tilde{z}(i_0 T) \end{bmatrix} - \int_0^T e^{\Lambda A \Lambda^{-1}(T-\tau)} \Lambda b \sigma(i_0 T + \tau) d\tau \tag{F.120}$$

Equation (F.119) can be written as:

$$\chi_{raw}((i_0 + 1)T) = \Upsilon(T) \begin{bmatrix} \tilde{y}(i_0 T) \\ 0_{(n-1)} \end{bmatrix} + \Phi(T)\hat{\sigma}(i_0 T) \tag{F.121}$$

With merging the adaptive law (F.24) into (F.121), the following is true:

First:

$$\chi_{raw}((i_0 + 1)T) = 0 + \Phi(T)\Phi^{-1}(T)1_1 h(i_0 T) \tag{F.122}$$

Second:

$$\zeta_{raw}((i_0+1)T) = \Upsilon(T)\begin{bmatrix} 0 \\ \tilde{z}(i_0 T) \end{bmatrix} - \int_0^T e^{\Lambda A \Lambda^{-1}(T-\tau)}\Lambda b\sigma(i_0 T + \tau)d\tau \qquad (\text{F.123})$$

The term $\Phi(T)\Phi^{-1}(T)1_1 h(i_0 T) = 1_1 h(i_0 T)$ from the equation for $\chi_{raw}((i_0+1)T)$ (where it originally belongs to because of $\hat{\sigma}(i_0 T)$) is now pulled into the equation of $\zeta_{raw}((i_0+1)T)$ in the form of $\Upsilon(T)\Upsilon^{-1}(T)1_1 h(i_0 T)$ such that $\zeta((i_0+1)T) = \zeta_{raw}((i_0+1)T) + \Upsilon(T)\Upsilon^{-1}(T)1_1 h(i_0 T)$. This is allowable because of Equation (F.117), in which a term from one summand is shifted to another summand – the sum is the same.

Hence:

$$\chi((i_0+1)T) = \chi_{raw}((i_0+1)T) - 1_1 h(i_0 T) = 0 \qquad (\text{F.124})$$

as well as:

$$\zeta((i_0+1)T) = \zeta_{raw}(i_0 T + t) + \Upsilon(T)\Upsilon^{-1}(T)\Phi(T)\Phi^{-1}(T)1_1 h(i_0 T) \qquad (\text{F.125})$$

Equation (F.125) is reformulated with the help of (F.123) to:

$$\zeta((i_0+1)T) = \Upsilon(T)\begin{bmatrix} 0 \\ \tilde{z}(i_0 T) \end{bmatrix} - \int_0^T e^{\Lambda A \Lambda^{-1}(T-\tau)}\Lambda b\sigma(i_0 T + \tau)d\tau + \Upsilon(T)\Upsilon^{-1}(T)1_1 h(i_0 T)$$
$$(\text{F.126})$$

In an adaptive law not featuring the additional term with $h(iT)$, equation (F.123) states the remaining term. This remaining term arises in the time span where the old cumulated uncertainties are canceled while new uncertainties are cumulating again. See section 2.5.2 for a more precise explanation. With the recursive definition of $h(iT)$ based on $\tilde{y}(iT)$, it can be seen that these remaining terms are taken into account via $h(iT)$. Note that $h(iT)$ is scalar:

$$h(iT) = -1_1^T e^{\Lambda A \Lambda^{-1}T}\begin{bmatrix} 0 \\ \tilde{z}'((i-1)T) \end{bmatrix}$$
$$+ 1_1^T \int_0^T e^{\Lambda A \Lambda^{-1}(T-\tau)}\Lambda b\sigma((i-1)T + \tau)d\tau + \nu_h(iT) \qquad (\text{F.127})$$

where $\tilde{z}'(t)$ is a hypothetical noise-free $\tilde{z}(t)$ to ensure uniqueness of noise in $\nu_h(iT)$.

Equation (F.126) can be rewritten to:

$$\zeta((i_0+1)T) = \Upsilon(T)\begin{bmatrix} v_1(i_0T) \\ v_2(i_0T) \end{bmatrix} - \int_0^T e^{\Lambda A\Lambda^{-1}(T-\tau)}\Lambda b\sigma(i_0T+\tau)d\tau \tag{F.128}$$

where $v_1(t)$ and $v_2(t)$ are defined in Equation (F.29).

Equation (F.128) is the solution of the following differential equation, where $t \in [i_0T, (i_0+1)T)$:

$$\dot{\zeta}(t) = \Lambda A\Lambda^{-1}\zeta(t) - \Lambda b\sigma(t) \tag{F.129}$$

with the initial condition of:

$$\zeta(i_0T) = \begin{bmatrix} v_1(i_0T) \\ v_2(i_0T) \end{bmatrix} = v(i_0T) \tag{F.130}$$

The positive definite function $V(t)$ was according to (F.40) defined as:

$$V(t) = \zeta^T(t)\Lambda^{-T}P\Lambda^{-1}\zeta(t) \tag{F.131}$$

With Λ being invertible (see Equation (F.18)) and $P > 0$ (see section F.5.3), it follows that $V > 0$ with $\Lambda^{-T}P\Lambda^{-1} > 0$.

The relocation of $h(i_0T)$ stated in Equations (F.124) and (F.125) in comparison to [13] is possible as the additional term $h(i_0T)$ is piece-wise constant, and therefore can be merged into the initial condition for a positive definite function $V(t)$ on the defined time length. This is true as the function $V(t)$ is used to examine the behavior in a continuous interval $t \in [i_0T, (i_0+1)T)$ over which $h(i_0T)$ is constant.

Applying Lemma F.3.1, for the positive definite function $V(t)$ at $t = i_0T$ the following is true:

$$V(i_0T) = k_1\zeta_1^2(i_0T) + \zeta_2^T(i_0T)K_2\zeta_2(i_0T) \tag{F.132}$$

With the assumption in (F.118) that $V(i_0T) \leq \alpha$ (needed to see whether this implies boundedness for $t = (i_0+1)T$ as well):

$$V(i_0T) \leq \alpha \tag{F.133}$$

For examining boundedness of $V(t)$ for $t \geq i_0 T$, the derivative is examined:

$$\dot{V}(t) = \zeta^T(t)\Lambda^{-T}P\Lambda^{-1}\dot{\zeta}(t) + \dot{\zeta}^T(t)\Lambda^{-T}P\Lambda^{-1}\zeta(t) \tag{F.134}$$

With Equation (F.129):

$$\dot{V}(t) = \zeta^T(t)\Lambda^{-T}P\Lambda^{-1}(\Lambda A\Lambda^{-1}\zeta(t) - \Lambda b\sigma) + (\Lambda A\Lambda^{-1}\zeta(t) - \Lambda b\sigma)^T\Lambda^{-T}P\Lambda^{-1}\zeta(t) \tag{F.135}$$

$$\dot{V}(t) = \zeta^T(t)\Lambda^{-T}(PA + A^TP)\Lambda^{-1}\zeta(t) - 2\zeta^T(t)\Lambda^{-T}Pb\sigma(t) \tag{F.136}$$

where Lemma F.1.3 can be applied:

$$PA + A^TP = -Q \tag{F.137}$$

At this point, it becomes obvious that two mutually dependent bounds appear. In section F.5.7 we see that an upper bound for $\sigma(t)$, namely Δ, is dependent on an upper bound for $\tilde{y}(t)$, namely $\bar{\gamma}_0$. For the positive definite function $V(t)$ an upper bound for $\sigma(t)$ is needed for finding an upper bound for $\tilde{y}(t)$. To avoid circular reasoning, the proof is conducted by contradiction. Starting in both (still to be proved) semiglobal positively invariant sets is trivial, since $\tilde{y}(0) \leq \varsigma$ (see Equation (F.102)). Now we hypothetically go to the border of the set for $\tilde{y}(t)$.

The hypothesis to be refuted is: "$|\tilde{y}(t)| < \bar{\gamma}_0$ is not always true". By starting with $\tilde{y}(0) = 0$, this hypothesis and a continuous $\tilde{y}(t)$ imply a time $0 < \iota$, where $|\tilde{y}(\iota)| = \bar{\gamma}_0$ and $|\tilde{y}(0 \leq t < \iota)| < \bar{\gamma}_0$.

This assumption is equivalent to the statement:

$$\|\tilde{y}_\iota\|_{\mathcal{L}_\infty} = \bar{\gamma}_0 \tag{F.138}$$

Lemma F.3.2 delivers with inverted signs for a general matrix A and a vector x:

$$-\lambda_{min}(A)\|x\|_2^2 \geq -x^TAx \geq -\lambda_{max}(A)\|x\|_2^2$$

Lemma F.4.1.

$$\text{If } \|\tilde{y}_\iota\|_{\mathcal{L}_\infty} = \bar{\gamma}_0, \text{ then } \|\sigma_\iota\|_{\mathcal{L}_\infty} \leq \Delta.$$

The proof can be found in section F.5.7.

Then:

$$\dot{V}(t) \leq -\lambda_{min}\left(\Lambda^{-T}Q\Lambda^{-1}\right)\|\zeta(t)\|_2^2 + 2\|\zeta(t)\|\left\|\Lambda^{-T}Pb\right\|_2\Delta \tag{F.139}$$

With $\Lambda^{-T}Q\Lambda^{-1}$ being positive definite (see explanation from Equation (F.131) where both, P and Q are positive definite, and the fact that a positive definite matrix always has only positive eigenvalues), the first term in Equation (F.139) can only be negative (including the $(-)$), whereas the second term is always positive.

Hence, the inequality $\dot{V}(t) < 0$ is fulfilled for:

$$\|\zeta\|_2 > \|\zeta_0\|_2 = \frac{2\left\|\Lambda^{-T}Pb\right\|_2\Delta}{\lambda_{min}\left(\Lambda^{-T}Q\Lambda^{-1}\right)} \tag{F.140}$$

Using this result into (F.131) provides with the help of Lemma F.3.2:

$$V(\zeta_0) = \zeta_0(t)^T\Lambda^{-T}P\Lambda^{-1}\zeta_0(t) \leq \lambda_{max}(\Lambda^{-T}P\Lambda^{-1})\left(\frac{2\left\|\Lambda^{-T}Pb\right\|_2\Delta}{\lambda_{min}\left(\Lambda^{-T}Q\Lambda^{-1}\right)}\right)^2 \leq \alpha \tag{F.141}$$

where α is defined in (F.38).

This and $V(i_0T) \leq \alpha$ show directly that:

$$V(t) \leq \alpha \quad \forall t \in [i_0T, (i_0+1)T), \quad (i_0+1)T \leq \wr \tag{F.142}$$

With $\chi(i_0T) = 0$ from Equation (F.124) and the fact that $\tilde{\xi}(t) = \zeta(t) + \chi(t)$ it follows that not only:

$$\zeta^T((i_0+1)T)\Lambda^{-T}P\Lambda^{-1}\zeta((i+1)T) \leq \alpha \tag{F.143}$$

but also:

$$\tilde{\xi}^T((i_0+1)T)\Lambda^{-T}P\Lambda^{-1}\tilde{\xi}((i+1)T) \leq \alpha \tag{F.144}$$

Using section F.5.5 with $\tilde{\xi}$ as the therein stated vector w, i.e.

$$\tilde{\xi}(i_0T) = \zeta(i_0T) = v(i_0T) \tag{F.145}$$

where the last equal sign is due to Equation (F.130):

$$\tilde{\xi}((i_0+1)T)^T \Lambda^{-T} P \Lambda^{-1} \tilde{\xi}((i_0+1)T) = k_1 v_1^2((i_0+1)T) + v_2^T((i_0+1)T) K_2 v_2((i_0+1)T) \leq \alpha \tag{F.146}$$

which means the upper bound from Equation (F.133) holds for $t = (i_0+1)T \leq \lambda$, too:

$$V((i_0+1)T) \leq \alpha \tag{F.147}$$

At this point it is proved that the second assumption of Equation (F.118), namely $V(i_0 T) \leq \alpha$ implies $V((i_0+1)T) \leq \alpha$, too. It is still to be proved that the same applies to the first assumption, namely from $\tilde{y}(i_0 T) \leq \varsigma$ implies $\tilde{y}((i_0+1)T) \leq \varsigma$. This is next:

From Equation (F.133) and with Lemma F.3.1 it follows that:

$$v^T(i_0 T)\Lambda^{-T} P \Lambda^{-1} v(i_0 T) \leq \alpha \tag{F.148}$$

This can be used with Lemma F.3.2 to obtain:

$$\lambda_{min}\left(\Lambda^{-T} P \Lambda^{-1}\right) \|v(i_0 T)\|_2^2 \leq v^T(i_0 T)\Lambda^{-T} P \Lambda^{-1} v(i_0 T) \tag{F.149}$$

Combining the last two equations delivers:

$$\lambda_{min}\left(\Lambda^{-T} P \Lambda^{-1}\right) \|v(i_0 T)\|_2^2 \leq \alpha \tag{F.150}$$

Which directly leads to:

$$\|v(i_0 T)\|_2 \leq \sqrt{\frac{\alpha}{\lambda_{min}\left(\Lambda^{-T} P \Lambda^{-1}\right)}} \tag{F.151}$$

Where:

$$\|v(i_0 T)\|_2 = \left\|\begin{bmatrix} v_1(i_0 T) \\ v_2(i_0 T) \end{bmatrix}\right\|_2 \tag{F.152}$$

meaning that $\begin{bmatrix} v_1(i_0 T), & v_2(i_0 T) \end{bmatrix}^T$ is bounded.

Next, as $\chi((i_0+1)T) = 0$:

$$\tilde{y}((i_0+1)T) = 1_1^T \zeta((i_0+1)T) + \nu_{rec}(i_0 T) \tag{F.153}$$

with the definition in (F.128):

$$\tilde{y}((i_0+1)T) = 1_1^T e^{\Lambda A\Lambda^{-1}T} \begin{bmatrix} v_1(i_0T) \\ v_2(i_0T) \end{bmatrix}' - 1_1^T \int_0^T e^{\Lambda A\Lambda^{-1}(T-\tau)} \Lambda b\sigma(i_0T + \tau)d\tau + \nu_{rec}((i_0+1)T)$$

(F.154)

where $v'(t)$ is a hypothetical noise-free $v(t)$ to ensure uniqueness of noise in $\nu_{rec}(iT)$. Note that this equation states the closed loop behavior, i.e. with $\hat{\sigma}(t)$ being applied.

An upper bound can be derived with Equation (F.150) as:

$$|\tilde{y}((i_0+1)T)| \leq \left\| 1_1^T \Upsilon(T) \right\|_2 \sqrt{\frac{\alpha}{\lambda_{min}\left(\Lambda^{-T}P\Lambda^{-1}\right)}}$$

(F.155)

$$+ \int_0^T \left| 1_1^T e^{\Lambda A\Lambda^{-1}(T-\tau)} \Lambda b \right| \left| \sigma(i_0T + \tau) \right| d\tau + \bar{\nu}_{rec}$$

$$\leq \varsigma$$

where according to Lemma F.4.1 $|\sigma(i_0T + \tau)| \leq \Delta$. Note that the assumptions from the proof of contradiction still hold. The boundedness of $v(t)$ is shown in (F.151), ς is defined in (F.39).

This means that from $\tilde{y}(i_0T) \leq \varsigma$ follows that $\tilde{y}((i+1)T) \leq \varsigma$ as well, where $(i_0+1)T \leq \wr$ and therefore $\forall iT \leq \wr$.

With (F.152) and from the fact that $k_1 > 0$ as well as $K_2 > 0$ ('positive definite') it follows that both, $v_1(i_0T)$ and $v_2(i_0T)$ are bounded. In the definition of $v(t)$ in Equation (F.29) there is no influence of $\tilde{z}(i_0T)$ to $v_1(i_0T)$. As $v_1(i_0T)$ is bounded and $h(i_0T)$ is scalar, it follows that $h(i_0T)$ is bounded as well. As the left side and the first summand of the equation are bounded, the second summand of the equation, namely $\tilde{z}(i_0T)$ must be bounded, too. This argumentation can be repeated for $t = (i_0+1)T$ as well by the validity of boundedness of $V(i_0T)$ and $y(i_0T)$ $\forall i_0 \in i$.

Third, is the proof of $\tilde{y}(iT + t) \leq \gamma_0$, where γ_0 is a constant and $0 \leq t \leq T$, $iT + t < \wr$. It is shown that a time $t = \wr$ where $\tilde{y}(\wr) = \bar{\gamma}_0$ is not reached.

Making use of Equation (F.114), one has for $t \in [0, T)$:

$$\tilde{y}(iT + t) = 1_1^T e^{\Lambda A\Lambda^{-1}t} \tilde{\xi}(iT) + 1_1^T \int_0^t e^{\Lambda A\Lambda^{-1}(t-\tau)} \Lambda \hat{\sigma}(iT)d\tau$$

(F.156)

$$- 1_1^T \int_0^t e^{\Lambda A\Lambda^{-1}(t-\tau)} \Lambda b\sigma(iT + \tau)d\tau + \nu_{rec}(iT)$$

Note that for the integral $\hat{\sigma}(iT)$ is a constant as $0 \leq t \leq iT$.

From Equation (F.145) it follows that $\tilde{\xi}(iT) = v(iT)$. Further, by using $\hat{\sigma}(iT)$ from the definition in (F.24) one has:

$$\tilde{y}(iT + t) = 1_1^T \Upsilon(t) v'(iT) + 1_1^T \Phi(t)(-\Phi^{-1}(T))\Upsilon(T)1_1 \tilde{y}'(iT) + \tag{F.157}$$
$$1_1^T \Phi(t)\Phi^{-1}(T)1_1 h'(iT) - 1_1^T \int_0^t e^{\Lambda A \Lambda^{-1}(t-\tau)} \Lambda b \sigma(iT + \tau)d\tau + \nu_{rec}(iT)$$

where $h'(t)$, $v'(t)$, $\tilde{y}'(t)$ and ς' are hypothetical noise-free $h(t)$, $v(t)$, $\tilde{y}(t)$ and ς to ensure uniqueness of noise in $\nu_{rec}(iT)$.

Now, with the extension to $t \in [0, T]$, $\tilde{y}(iT + t)$ is bounded by some constant γ_0, i.e. $\tilde{y}(iT + t) \leq \gamma_0$:

$$\|\tilde{y}(iT + t)\|_\infty \leq \max_{t \in [0,T]} \left(1_1^T e^{\Lambda A \Lambda^{-1}t}\right) \sqrt{\frac{\alpha}{\lambda_{min}\left(\Lambda^{-T} P \Lambda^{-1}\right)}}$$
$$+ \max_{t \in [0,T]} \left(\int_0^t \left|1_1^T e^{\Lambda A \Lambda^{-1}(t-\tau)}\Lambda \Phi^{-1}(T)\Upsilon(T)1_1\right| d\tau\right) \varsigma' \tag{F.158}$$
$$+ \max_{t \in [0,T]} \left(\int_0^t \left|1_1^T e^{\Lambda A \Lambda^{-1}(t-\tau)}\Lambda \Phi^{-1}(T)1_1\right| d\tau\right) \|h'(iT)\|_2$$
$$+ \max_{t \in [0,T]} \left(\int_0^t \left|1_1^T e^{\Lambda A \Lambda^{-1}(t-\tau)}\Lambda b\right| d\tau\right) \Delta + \bar{\nu}_{rec}$$
$$= \gamma_0$$

It is shown on page 156 that $\|h(iT)\|_2$ exists.

Lemma F.4.2.
$$\lim_{T \to 0} \gamma_0 = \nu_{rec}(t)$$

The proof is shown in section F.5.8.

Remark F.4.1.

In Equation (F.157) some terms show t, others T. To overcome assorted dependencies, the maximum in the span $t \in [0, T]$ is taken.

Remark F.4.2.

For helicopters, considering times between iT and $(i+1)T$ is meaningless. The step length T is much shorter than relevant dynamics, hence if the signals are bounded at iT and $(i+1)T$, they are as well bounded at times between and cannot change significantly.

For completing the contradiction, it must be true that $\tilde{y}(iT + t) < \bar{\gamma}_0$ which implies that $\gamma_0 < \bar{\gamma}_0$ has to hold. The value of $\bar{\gamma}_0$ was chosen beforehand as some guessed value and first appears in the assumption of (F.138). With $\lim_{T \to 0} \gamma_0 = \nu_{rec}(t)$ it is clear that for every $\bar{\gamma}_0$ a limit of $\tilde{y}(iT + t)$, namely γ_0 can be made arbitrarily small except for limits of sensor noise, if only the computation step length T is chosen small enough. This rounds off the proof via contradiction as the hypothesis in (F.138) is refuted by this.

By this method, circular reasoning is avoided. This would occur if T is chosen and the performance bounds are tried to be found. By iterating this procedure – choosing $\bar{\gamma}_0$ and finding the necessary maximum T, the design can be theoretically done by first stating the performance bounds in terms of $\bar{\gamma}_0$ and finding the required CPU performance. With this, it must not be forgotten that the bounds are found with very conservative elements in the proof and is therefore not practical for most applications.

Next, $\|(y_{ref} - y)_i\|_{\mathcal{L}_\infty}$ and $\|(u_{ref} - u)_i\|_{\mathcal{L}_\infty}$ are analyzed:

From Equation (F.230) and with $\|\tilde{y}_i\| < \bar{\gamma}_0$ one has:

$$\|e_i\|_{\mathcal{L}_\infty} = \|(y_{ref} - y)_i\|_{\mathcal{L}_\infty} < \frac{\|H_2(s)\|_{\mathcal{L}_1}}{1 - \|H(s)(1 - C(s))\|_{\mathcal{L}_1} L} \bar{\gamma}_0 = \gamma_1 \qquad \text{(F.159)}$$

Likewise, $\|(u_{ref} - u)_i\|_{\mathcal{L}_\infty}$ is bounded by a constant γ_2.

Lemma F.4.3.

$$\|(u_{ref} - u)_i\|_{\mathcal{L}_\infty} < \gamma_2$$

The proof can be found in section F.5.9.

At this point, all bounds (F.109), (F.110), (F.111) are found.

Remark F.4.3.

In all simulations conducted, it is found that the first component of $\Upsilon^{-1}(T)$ is significantly greater (while being close to 1) than the other components in the first column (those being about in the order of magnitude of $T/10...T$). This means that $[v_1(iT), v_2(iT)]^T$ in these cases is very close to $[h(iT), \tilde{z}(iT)]^T$. Then $h(iT)$ has very little influence on $\tilde{z}(t)$ but significant influence on $\tilde{y}(t)$.

This observation is consistent with (F.202), i.e. $\lim_{T \to 0} \Upsilon(T) = I_{(n \times n)}$.

Remark F.4.4.

With this adaptive law it should be possible to find tighter theoretical performance bounds or even convergence of $\tilde{y}(t) \to 0$ (without noise) as it is done in [18]. This however is omitted here as a revised proof in this direction does not change the controller performance. The necessity of not "blowing up" however is shown here and therefore regarded sufficient as the bounds will not be verified as being physically meaningful.

Remark F.4.5.

The function $V(t)$ is positive definite rather than positive semidefinite due to the positive definiteness of the matrices Q and therefore also P in the algebraic Lyapunov equation from Lemma F.1.3.

Remark F.4.6.

A scalar system is not meaningful in this approach. The minimum order of $M(s)$ is therefore 2. In case of a scalar system many parts of the proof become meaningless, e.g. the nullspace D is empty.

The requirement for a stable $H(s)$ by the choice of $M(s)$ and $C(s)$ can be examined on its robustness. The fact that stability of $H(s) = \frac{F(s)M(s)}{C(s)F(s)+(1-C(s))M(s)}$ and the one of stabilization of $F(s)$ by $\frac{C(s)}{M(s)(1-C(s))}$ are identical [13], allows for evaluating a chosen $C(s)$ regarding its stability effect on $H(s)$ for a family of transfer functions $F_\delta(s)$. E.g. a nominal $F_0(s) = \frac{s+4}{s^2+8s+15}$ can tolerate about 89% uncertainty in the nominal parameters k_i of $\frac{(s+k_1)}{(s+k_2)(s+k_3)}$ until instability, where $M(s) = 2\frac{s+20}{s^2+25s+150}$ and $C(s) = \frac{1}{1/20s+1} \cdot \frac{1}{1/(1.2\cdot20)s+1}$. This seems an useful robustness evaluation of possible plant properties at large, but it refers to linear systems with parametric uncertainties – a fact because of which these considerations are not further pursued in this thesis. Instead, the experimentally obtained time delay margins are regarded a paramount measurement of robustness. Both methods are of course not mutually exclusive.

In short, the following design constraints apply:

- In augmentation $M(s)$ must only show what the baseline controller accounts for.

- $reldeg(M(s)) = 1$.

- $M(s)$ must be chosen stable without positive zeros.

- Zeros and poles of $M(s)$ must be sufficiently far away from zero to avoid lightly damped poles and zeros, which evoke peaking phenomena. Note that $M(s)$ is inverted in the control law.

- $\mathcal{L}_1$-norm condition from (F.14) must hold.

- $C(s)$ and $M(s)$ must be chosen for $H(s)$ to be stable.

- DC-gain of $C(s)$ must be 1 for disturbance rejection with zero steady state error.

- $M(s)$ and $C(s)$ should not be more complicated than necessary to avoid overly high CPU loads.

Summary of inclusion of the measurement noise:

By injecting measurement noise into the predictor, dynamic effects occur, namely the current noise value in $\tilde{y}(iT)$ is dependent on the current noise and its entire history, the noise enters the prediction error through a virtual discrete high-pass filter. Approximating discrete behavior by continuous transfer functions allows for a correct evaluation of low-frequency noise, high frequencies however are not correctly described by it. For the latter, the current addition of every time step dominates the effects from history. Within the time span of $t \in [iT, (i+1)T]$ a more accurate description is the addition of new noise $\nu((i+1)T)$ every $t = (i+1)T$, meaning that $\nu(iT)$ is somehow included in the approximation error $\varepsilon(t)$.

The measurement noise is primarily a degradation of the adaption quality expressed in $\tilde{y}(t)$, which is then propagated to the errors $y_{ref}(t) - y(t)$ and $u_{ref}(t) - u(t)$.

F.5 Proof of Lemmas

F.5.1 Proof of Lemma F.1.1

The relative degree of $M(s)$ is chosen according to the relative degree of the transfer functions $H(s), H_0(s), H_1(s), H_2(s), H_3(s)$. Only proper – preferably strictly proper for noise attenuation – is acceptable.

General Framework:

According to the nomenclature, define numerator and denominator of a transfer function as:

$$C(s) = \frac{C_n(s)}{C_d(s)},\ F(s) = \frac{F_n(s)}{F_d(s)},\ M(s) = \frac{M_n(s)}{M_d(s)}, \quad \text{etc. and } \wp(h(s)) \text{ denotes the order of } h(s).$$

Assume two transfer functions D(s) and G(s), then:

Proposition 1: Addition

$$D(s) + G(s) = \frac{D_n(s)}{D_d(s)} + \frac{G_n(s)}{G_d(s)} = \frac{D_n(s)G_d(s) + G_n(s)D_d(s)}{D_d(s)G_d(s)} \tag{F.160}$$

$$\frac{\wp(D(s) + G(s))_n}{\wp(D(s) + G(s))_d} = \frac{max\left(\wp(D_n(s)) + \wp(G_d(s)),\, \wp(G_n(s)) + \wp(D_d(s))\right)}{\wp(D_d(s)) + \wp(G_d(s))} \tag{F.161}$$

Proposition 2: Multiplication

$$D(s)G(s) = \frac{D_n(s)}{D_d(s)}\frac{G_n(s)}{G_d(s)} \tag{F.162}$$

$$\frac{\wp(D(s)G(s))_n}{\wp(D(s)G(s))_d} = \frac{\wp(D_n(s)) + \wp(G_n(s))}{\wp(D_d(s)) + \wp(G_d(s))} \tag{F.163}$$

Proposition 3: Expandability

$$reldeg\left(\frac{h(s)G_n(s)}{h(s)G_d(s)}\right) = reldeg\left(\frac{G_n(s)}{G_d(s)}\right) \tag{F.164}$$

where $h(s)$ is a polynomial.

Proof.

$$reldeg\left(\frac{h(s)G_n(s)}{h(s)G_d(s)}\right) = \wp(h(s)G_d(s)) - \wp(h(s)G_n(s))$$
$$= \wp(h(s)) + \wp(G_d(s)) - (\wp(h(s)) + \wp(G_n(s)))$$
$$= \wp(G_d(s)) - \wp(G_n(s))$$
$$= reldeg(G(s))$$

$\square$

1. $H(s)$:

$$H(s) = \frac{F_n(s)C_d(s)M_n(s)}{F_n(s)C_n(s)M_d(s) + F_d(s)(C_d(s) - C_n(s))M_n(s)} \tag{F.165}$$

With $C(s)$ being strictly proper, i.e. $\wp(C_d(s)) > \wp(C_n(s))$, it follows that $\wp(C_d(s) - C_n(s)) = \wp(C_d(s))$ according to Proposition 3.

It follows from Proposition 2 that:

$$\frac{\wp(H(s))_n}{\wp(H(s))_d} = \frac{\wp(F_n(s)) + \wp(C_d(s)) + \wp(M_n(s))}{max(\wp(F_n(s)) + \wp(C_n(s)) + \wp(M_d(s)),\ \wp(F_d(s)) + \wp(C_d(s)) + \wp(M_n(s)))} \tag{F.166}$$

It is sufficient to show that one of the arguments of $max(\cdot, \cdot)$ in the denominator of Equation (F.166) leads to a strictly proper $H(s)$. If this term is equal to the maximum of both arguments this case is trivial, if the first argument to be tried shows higher order than the second one, this is on the safe side in the proof for $H(s)$ to be strictly proper.

Applying Proposition 3 to the second argument with the fact that $\wp(F_d(s)) > \wp(F_n(s))$ proofs that $H(s)$ is strictly proper, i.e. $\wp(H_d(s)) - \wp(H_n(s)) > 0$.

Hence $reldeg(H(s)) \geq reldeg(F(s))$.

As $H(s)$ is strictly proper, $H(s)(1 - C(s))$ is strictly proper, too.

2. $H_0(s)$

The critical transfer function is $H_0(s)$ which shows the following behavior:

$reldeg(F(s)) < reldeg(M(s)) \quad \rightarrow H_0$ improper
$reldeg(F(s)) = reldeg(M(s)) \quad \rightarrow H_0$ proper, not strictly proper
$reldeg(F(s)) > reldeg(M(s)) \quad \rightarrow H_0$ strictly proper

A lower bound for the relative degree can be stated by assuming that no direct feedthrough exists: $reldeg(F(s)) \geq 1$.

According to this, $reldeg(M(s))$ must be chosen 0 or 1. A transfer function with relative degree 0 implicates direct feedthrough, which does not match the physics of a helicopter. This is why in the implementation only a relative degree of 1 qualifies.

Proof.

$$H_0(s) = \frac{F_n(s)C_d(s)M_d(s)}{F_n(s)C_n(s)M_d(s) + F_d(s)(C_d(s) - C_n(s))M_n(s)} \tag{F.167}$$

With $C(s)$ being strictly proper, i.e. $\wp(C_d(s)) > \wp(C_n(s))$, it follows that
$\wp(C_n(s) - C_d(s)) = \wp(C_d(s))$. From this, we can write:

$$\frac{\wp(H_0(s))_n}{\wp(H_0(s))_d} = \frac{\wp(F_n(s)) + \wp(C_d(s)) + \wp(M_d(s))}{max(\wp(F_n(s)) + \wp(C_n(s)) + \wp(M_d(s)),\ \wp(F_d(s)) + \wp(C_d(s)) + \wp(M_n(s)))} \tag{F.168}$$

The first argument of the $max(\cdot, \cdot)$ operator would imply an improper $H_0(s)$ according to
Proposition 3. The second term delivers with the help of Proposition 3:

$$\begin{aligned}
reldeg(H_0(s)) &= \wp(F_d(s)) - \wp(F_n(s)) + \wp(M_n(s)) - \wp(M_d(s)) \\
&= reldeg(F(s)) - reldeg(M(s))
\end{aligned} \tag{F.169}$$

$\square$

3. $H_1(s)$

$$H_1(s) = \frac{F_n(s)C_n(s)M_d(s) - F_d(s)C_n(s)M_n(s)}{F_n(s)C_n(s)M_d(s) + (C_d(s) - C_n(s))F_d(s)(s)M_n(s)} \tag{F.170}$$

With $C(s)$ being strictly proper, i.e. $\wp(C_d(s)) > \wp(C_n(s))$, it follows that
$\wp(C_n(s) - C_d(s)) = \wp(C_d(s))$. Then:

$$\frac{\wp(H_1(s))_n}{\wp(H_1(s))_d} = $$

$$\frac{max\left(\wp(F_n(s)) + \wp(C_n(s)) + \wp(M_d(s)),\ \wp(F_d(s)) + \wp(C_n(s)) + \wp(M_n(s))\right)}{max\left(\wp(F_n(s)) + \wp(C_n(s)) + \wp(M_d(s)),\ \wp(C_d(s)) + \wp(F_d(s)) + \wp(M_n(s))\right)} \tag{F.171}$$

The second argument in the numerator is greater than the first one.

Proof.
$reldeg(F(s)) \geq reldeg(M(s))$ from evaluation of $H_0(s)$ leads to:

$$\wp(F_d(s)) - \wp(F_n(s)) \geq \wp(M_d(s)) - \wp(M_n(s))$$

$$\wp(F_d(s)) + \wp(C_n(s)) + \wp(M_n(s)) \geq \wp(F_n(s)) + \wp(C_n(s)) + \wp(M_d(s))$$

$\square$

As the first term of the numerator and the first term of denominator are equal, it remains to compare the respective second terms.

It can be seen with Proposition 3 that $reldeg(H_1(s)) = reldeg(C(s))$.

4. $H_2(s)$

As $F(s)$ is assumed to have $reldeg(F(s)) \geq 1$, the examination of $H(s)$ shows that $H(s)$ has at least the same relative degree $reldeg(H(s)) \geq 1$. In the same section, it is stated that $reldeg(M(s)) = 1$, meaning that $\frac{H(s)}{M(s)}$ has a relative degree of $reldeg(H(s)/M(s)) \geq 0$. The relative degree of $H_2(s)$ is further determined by $C(s)$ with a minimum relative degree of 1. This implies that $H_2(s)$ is strictly proper.

5. $H_3(s)$

$$H_3(s) = H_0(s)\frac{M(s)C(s)}{F(s)} \tag{F.172}$$

$$reldeg(H_3(s)) = reldeg(H_0(s)) + reldeg\left(\frac{M(s)C(s)}{F(s)}\right) \tag{F.173}$$

From the evaluation of $H_0(s)$ it follows that:

$$reldeg(H_0(s)) = reldeg(F(s)) - reldeg(M(s)) \tag{F.174}$$

$$reldeg\left(\frac{M(s)C(s)}{F(s)}\right) = \wp(F_n(s)) + \wp(C_d(s)) + \wp(M_d(s)) \\ - (\wp(F_d(s)) + \wp(C_n(s)) + \wp(M_n(s))) \tag{F.175}$$

Combining the last two equations implies:

$$\begin{aligned} reldeg(H_3(s)) &= \wp(F_d(s)) - \wp(F_n(s)) - (\wp(M_d(s)) - \wp(M_n(s))) + \\ &\quad \wp(F_n(s)) + \wp(C_d(s)) + \wp(M_d(s)) - (\wp(F_d(s)) + \wp(C_n(s)) + \wp(M_n(s))) \\ &= \wp(C_d(s)) - \wp(C_n(s)) \\ &= reldeg(C(s)) \end{aligned} \tag{F.176}$$

$\square$

F.5.2   Proof of Lemma F.1.2

If $C(s)$ and $M(s)$ are chosen such that $H(s)$ is stable, then $H_0(s), H_1(s), H_3(s)$ are stable as well since they feature the same denominator. With $M(s)$ and $C(s)$ being chosen stable, it follows that $H_2(s)$ is stable.

$\square$

F.5.3   Proof of Lemma F.1.3

Assume a system $\dot{x}(t) = Ax(t)$ with A being Hurwitz. Further, a matrix $Q = Q^T > 0$ is chosen which determines a matrix $P > 0$ such that $A^T P + PA = -Q$. Now consider a Lyapunov function candidate $V(t) = \frac{1}{2} x^T(t) P x(t)$. Its negative derivative implies asymptotic stability,
$\dot{V}(t) = \dot{x}(t)^T P x(t) + x^T(t) P \dot{x}(t) < 0$. Applying the system equation gives $x^T(t) \left(A^T P + PA \right) x(t) = -x^T(t) Q x(t) < 0$.

$\square$

F.5.4   Proof of Lemma F.2.1

(F.49) in (F.48) delivers:

$$u_{ref}(s) = \frac{M(s)C_r(s)r(s) - C(s)(F(s) - M(s))u_{ref}(s) - C(s)F(s)d_{ref}(s)}{M(s)}$$

$$u_{ref}(s) = \frac{M(s)C_r(s)r(s) - C(s)F(s)d_{ref}(s)}{M(s) + C(s)(F(s) - M(s))}$$

$$u_{ref}(s) = \frac{M(s)C_r(s)r(s) - C(s)F(s)d_{ref}(s)}{M(s)(1 - C(s)) + C(s)F(s)} \tag{F.177}$$

From (F.47) we have:

$$y_{ref}(s) = M(s)(u_{ref}(s) + \sigma_{ref}(s)) + y_{in}(s) \tag{F.178}$$

with (F.49):

$$y_{ref}(s) = M(s)\left(u_{ref}(s) + \frac{(F(s) - M(s))u_{ref}(s) + F(s)d_{ref}(s)}{M(s)}\right) + y_{in}(s) \qquad \text{(F.179)}$$

$$y_{ref}(s) = F(s)u_{ref}(s) + F(s)d_{ref}(s) + y_{in}(s) \qquad \text{(F.180)}$$

$$y_{ref}(s) = \frac{F(s)(M(s)C_r(s)r(s) - C(s)F(s)d_{ref}(s))}{M(s)(1 - C(s)) + C(s)F(s)} + F(s)d_{ref}(s) + y_{in}(s) \qquad \text{(F.181)}$$

$$y_{ref}(s) = H(s)C_r(s)r(s) + H(s)(1 - C(s))d_{ref}(s) + y_{in}(s) \qquad \text{(F.182)}$$

As $C(s)$ and $M(s)$ must be selected so that $H(s)$ is stable and strictly proper, one has:

$$\|y_{ref}\|_{\mathcal{L}_\infty} \leq \|H(s)C_r(s)\|_{\mathcal{L}_1} \|r\|_{\mathcal{L}_\infty} + \|H(s)(1 - C(s))\|_{\mathcal{L}_1} \|d_{ref}\|_{\mathcal{L}_\infty} + \|y_{in}\|_{\mathcal{L}_\infty}$$

$$\|y_{ref}\|_{\mathcal{L}_\infty} \leq \|H(s)C_r(s)\|_{\mathcal{L}_1} \|r\|_{\mathcal{L}_\infty} + \|H(s)(1 - C(s))\|_{\mathcal{L}_1} \left(L \|y_{ref}\|_{\mathcal{L}_\infty} + L_0\right) + \|y_{in}\|_{\mathcal{L}_\infty}$$

Solving this equation and implying the $\mathcal{L}_1$-norm condition (F.14):

$$\|y_{ref,\imath}\|_{\mathcal{L}_\infty} \leq \rho_r = \frac{\|H(s)C_r(s)\|_{\mathcal{L}_1} \|r\|_{\mathcal{L}_\infty} + \|H(s)(1 - C(s))\|_{\mathcal{L}_1} L_0 + \|y_{in}\|_{\mathcal{L}_\infty}}{1 - \|H(s)(1 - C(s))\|_{\mathcal{L}_1} L} < \infty$$

$$\text{(F.183)}$$

where $\|y_{in}\|_{\mathcal{L}_\infty}$ is bounded by ρ_{x_0} and c^T.

Remark F.5.1.

An intuitive explanation is shown on page 27 with Equation (2.12) and Equation (2.13), why the $\mathcal{L}_1$-norm condition benefits from high low-pass bandwidth.

$\square$

F.5.5 Proof of Lemma F.3.1

$$w^T \Lambda^{-T} P \Lambda^{-1} w = w^T \left(\sqrt{P} \Lambda^{-1} \right)^T \left(\sqrt{P} \Lambda^{-1} \right) w \tag{F.184}$$

The right-hand side of the last equation can be formulated as:

$$w^T \left(\sqrt{P} \Lambda^{-1} \right)^T \left(\sqrt{P} \Lambda^{-1} \right) w = w^T \left(\left(\Lambda \sqrt{P}^{-1} \right) \left(\Lambda \sqrt{P}^{-1} \right)^T \right)^{-1} w \tag{F.185}$$

$$
\begin{aligned}
\left(\Lambda \sqrt{P}^{-1} \right) \left(\Lambda \sqrt{P}^{-1} \right)^T &= \begin{bmatrix} c^T \sqrt{P}^{-1} \\ D \end{bmatrix} \left[\left(c^T \sqrt{P}^{-1} \right)^T \quad D^T \right] \\
&= \begin{bmatrix} c^T \sqrt{P}^{-1} \left(c^T \sqrt{P}^{-1} \right)^T & c^T \sqrt{P}^{-1} D^T \\ D \left(c^T \sqrt{P}^{-1} \right)^T & D D^T \end{bmatrix} \\
&= \begin{bmatrix} g_1 & 0 \\ 0 & G_2 \end{bmatrix}
\end{aligned} \tag{F.186}
$$

The non-diagonal entries of the last matrix vanish due to the choice of D to define the left nullspace of $\left(c^T \sqrt{P}^{-1} \right)^T$.

Both, λ and $\sqrt{P}$ are nonsingular, therefore G_2 is invertible.

By defining g_1^{-1} as k_1 and G_2^{-1} as K_2, it follows that

$$\left(\left(\Lambda \sqrt{P}^{-1} \right) \left(\Lambda \sqrt{P}^{-1} \right)^T \right)^{-1} = \begin{bmatrix} k_1 & 0 \\ 0 & K_2 \end{bmatrix} \tag{F.187}$$

The left-hand side of the last equation coincides with the left-hand side of Equation (F.184). The right-hand side can be verified by computing $w^T \begin{bmatrix} k_1 & 0 \\ 0 & K_2 \end{bmatrix} w$.

Remark F.5.2.

The choice of Λ becomes obvious now. The first component is to ensure that the output coincides with the first component of the transformed state vector. The nullspace is chosen for setting all non-diagonal entries to zero in Equation (F.186), whereas the inclusion of the matrix P from the algebraic Lyapunov equation serves as positive definite element in the positive definite function $V(t)$.

$\square$

F.5.6   Proof of Lemma F.3.2

Let L be a matrix with the eigenvalues λ_i on its diagonal and E be the matrix of eigenvectors $[e_1, e_2, ..., e_n]$ of the matrix $A \in \mathbb{R}^{n \times n}$.

The definition of eigenvalues and eigenvectors implies $AE = EL$.

This means with $E^T E = I$, A can be decomposed as $A = E^T LE$:

$$x^T A x = x^T E^T L E x = (Ex)^T L E x = \eta^T L \eta \tag{F.188}$$

where $\eta = Ex$. As L is diagonal we have:

$$\eta^T L \eta = \sum_{i=1}^{n} \lambda_i (\eta^T \eta)_i \tag{F.189}$$

Considering that:

$$(\eta^T \eta)_i = ((Ex)^T Ex)_i = (x^T E^T Ex)_i = (x^T x)_i \tag{F.190}$$

Then:

$$\eta^T L \eta = \sum_{i=1}^{n} \lambda_i x_i^2 = x^T A x \tag{F.191}$$

From this and the fact that:

$$\|x\|_2^2 = \sum_{i=1}^{n} x_i^2 \tag{F.192}$$

Lemma F.3.2 follows directly.

$\square$

F.5.7   Proof of Lemma F.4.1

$$y(t) = y_{ref}(t) - e(t) \tag{F.193}$$

$$\|y_i\|_{\mathcal{L}_\infty} \leq \|y_{ref,i}\|_{\mathcal{L}_\infty} + \|e_i\|_{\mathcal{L}_\infty} \tag{F.194}$$

where the index ι signals the truncated norm.

An upper bound for the first term of the right hand side can be found in section F.2:
$\|y_{ref,\iota}\|_{\mathcal{L}_\infty} \leq \rho_r$.

Lemma F.5.1.
$$\|e_\iota\|_{\mathcal{L}_\infty} \leq \frac{\|H_2(s)\|_{\mathcal{L}_1}}{1 - \|(H(s)(1 - C(s)))\|_{\mathcal{L}_1} L} \|\tilde{y}\|_{\mathcal{L}_\infty}$$

The proof can be found in section F.5.10.

This means:
$$\|y_\iota\|_{\mathcal{L}_\infty} \leq \rho_r + \frac{\|H_2(s)\|_{\mathcal{L}_1}}{1 - \|(H(s)(1 - C(s)))\|_{\mathcal{L}_1} L} \|\tilde{y}\|_{\mathcal{L}_\infty} \tag{F.195}$$

where $H_2(s)$ is defined in (F.12).

From Equation (F.224) we have:

$$\sigma(s) = \frac{(F(s) - M(s))(C_r(s)r(s) - \tilde{\sigma}(s)) + F(s)d(s)}{M(s)(1 - C(s)) + C(s)F(s)} \tag{F.196}$$

This, together with Equation (F.234), namely $\tilde{y}(s) = \frac{M(s)}{C(s)}\tilde{\sigma}(s)$:

$$\sigma(s) = \frac{(F(s) - M(s))(C_r(s)r(s) - \frac{C(s)}{M(s)}\tilde{y}(s)) + F(s)d(s)}{M(s)(1 - C(s)) + C(s)F(s)} \tag{F.197}$$

$$\sigma(s) = H_{1,r}(s)r(s) + \frac{H_1(s)}{M(s)}\tilde{y}(s) + H_0(s)d(s) \tag{F.198}$$

where $H_1(s)$ is defined in (F.10) and $H_{1,r}(s)$ in (F.11) and $H_0(s)$ in (F.9).

Due to the fact that $H_1(s), \frac{H_1(s)}{M(s)}$, and $H_0(s)$ are strictly proper and bounded:

$$\|\sigma_\iota\|_{\mathcal{L}_\infty} \leq \|H_{1,r}(s)\|_{\mathcal{L}_1} \|r\|_{\mathcal{L}_\infty} + \left\|\frac{H_1(s)}{M(s)}\right\|_{\mathcal{L}_1} \|\tilde{y}_\iota\|_{\mathcal{L}_\infty} + \|H_0(s)\|_{\mathcal{L}_1}\left(L \|y_\iota\|_{\mathcal{L}_\infty} + L_0\right) \tag{F.199}$$

With both, $H_{1,r}(s)$ and $M(s)$ being proper and stable, the $\mathcal{L}_1$-norm of $\frac{H_1(s)}{M(s)}$ is guaranteed to exist.

Now, the relation from Equation (F.195) can be applied to gain:

$$\|\sigma_l\|_{\mathcal{L}_\infty} \leq \|H_{1,r}(s)\|_{\mathcal{L}_1}\|r\|_{\mathcal{L}_\infty} + \|H_0(s)\|_{\mathcal{L}_1}(L\rho_r + L_0)$$
$$+ \left(\left\|\frac{H_1(s)}{M(s)}\right\|_{\mathcal{L}_1} + \|H_0(s)\|_{\mathcal{L}_1}L\frac{\|H_2(s)\|_{\mathcal{L}_1}}{1 - \|H(s)(1 - C(s))\|_{\mathcal{L}_1}L}\right)\|\tilde{y}_l\|_{\mathcal{L}_\infty} \tag{F.200}$$

Note that if $\|\tilde{y}_l\|_{\mathcal{L}_\infty} = \bar{\gamma}_0$, then $\|\sigma_l\|_{\mathcal{L}_\infty} \leq \Delta$, where Δ is defined in (F.36).

$\square$

F.5.8 Proof of Lemma F.4.2 and Lemma F.3.3

The proofs of the two lemmas coincide, thus it is exemplarily done for Lemma F.4.2.

Equation (F.157) embraces four main terms (summands) whose limit 0 for $T \to 0$ is proved so that $\tilde{y}(iT + t)$, $t \in [0, T)$ converges to a bounded noise term.

3rd **term:**

The third term is: $1_1^T \Phi(t)\Phi^{-1}(T)1_1 h'(iT)$

where according to Equation (F.127) but without noise:

$$h'(iT) = -1_1^T \Upsilon(T)\begin{bmatrix} 0 \\ \tilde{z}'((i-1)T) \end{bmatrix} + 1_1^T \int_0^T e^{\Lambda A \Lambda^{-1}(T-\tau)}\Lambda b\sigma((i-1)T + \tau)d\tau \tag{F.201}$$

$$\lim_{T \to 0} \Upsilon(T) = \mathbb{I} \tag{F.202}$$

Hence:

$$\lim_{T \to 0} 1_1^T \Upsilon(T) = 1_1^T \tag{F.203}$$

and:

$$\lim_{T \to 0} 1_1^T \Upsilon(T)\begin{bmatrix} 0 \\ \tilde{z}'((i-1)T) \end{bmatrix} = 0 \tag{F.204}$$

Together with:

$$\lim_{T \to 0} 1_1^T \int_0^T e^{\Lambda A \Lambda^{-1}(T-\tau)}\Lambda b\sigma((i-1)T + \tau)d\tau = 0 \tag{F.205}$$

It follows that:

$$\lim_{T \to 0} h'(iT) = 0 \tag{F.206}$$

With $\int_0^t \left| 1_1^T e^{\Lambda A \Lambda^{-1}(t-\tau)} \Lambda \Phi^{-1}(T) 1_1 \right| d\tau$ being bounded the third term is proved of $\lim_{T \to 0}(3rd\ term) = 0$.

1st **term:**

The first term is: $1_1^T \Upsilon(t) v'(iT)$

The definition of $v(iT)$ from Equation (F.29) is:

$$v'(iT) = \begin{bmatrix} v_1(iT) \\ v_2(iT) \end{bmatrix}' = \Upsilon^{-1}(T) 1_1 h'(iT) + \begin{bmatrix} 0 \\ \tilde{z}'(iT) \end{bmatrix} \tag{F.207}$$

With $\lim_{T \to 0} h'(iT) = 0$ and $\lim_{T \to 0} \left(\max_{t \in [0,T]} (\Upsilon(t)) \right) = I_{(n \times n)}$:

$$\lim_{T \to 0} 1_1^T \left(\max_{t \in [0,T]} \Upsilon(t) \right) v'(iT) = 1_1^T \begin{bmatrix} 0 \\ \tilde{z}'(iT) \end{bmatrix} = 0 \tag{F.208}$$

which completes the proof for the first term.

2nd **term:**

The second term is: $1_1^T \Phi(t)(-\Phi^{-1}(T)) \Upsilon(T) 1_1 \tilde{y}'(iT)$.

With the previous arguments, it can be verified with (F.154) that:

$$\lim_{T \to 0} \tilde{y}'(iT) = 0 \tag{F.209}$$

With $\int_0^t \left| 1_1^T e^{\Lambda A \Lambda^{-1}(t-\tau)} \Phi^{-1}(T) \Upsilon(T) 1_1 \right| d\tau$ being bounded, it follows the limit of the second term being 0.

4th **term:**

The fourth term is: $-1_1^T \int_0^t e^{\Lambda A \Lambda^{-1}(t-\tau)} \Lambda b \sigma(iT + \tau) d\tau$

Similar to the proofs above, one finds that:

$$\lim_{T \to 0} \left(\max_{t \in [0,T]} \int_0^t \left| 1_1^T e^{\Lambda A \Lambda^{-1}(t-\tau)} \Lambda b \right| d\tau \right) = 0 \tag{F.210}$$

Together with the assumption $\tilde{y}(\imath) = \bar{\gamma}_0$ from (F.138), meaning with $\|\sigma_\imath\|_{\mathcal{L}_\infty} \leq \Delta$ from section F.5.7, the fourth's term limit is proved, too.

The term $\nu_{rec}(t)$ is remaining and cannot be avoided.

$\square$

F.5.9   Proof of Lemma F.4.3

Let $\tilde{\sigma}(s)$ be defined as:

$$\tilde{\sigma}(s) := \frac{C(s)}{M(s)} c^T \left(s\mathbb{I} - A \right)^{-1} \hat{\sigma}(s) - C(s)\sigma(s) \tag{F.211}$$

According to Equation (F.28), the control signal is defined by:

$$u(s) = C_r(s)r(s) - \frac{C(s)}{M(s)} c^T \left(s\mathbb{I} - A \right)^{-1} \hat{\sigma}(s) \tag{F.212}$$

Then with (F.211):

$$u(s) = C_r(s)r(s) - \tilde{\sigma}(s) - C(s)\sigma(s) \tag{F.213}$$

Using the definition of $\sigma(s)$ from (F.6) in (F.213):

$$u(s) = C_r(s)r(s) - \tilde{\sigma}(s) - \frac{C(s)(F(s) - M(s))u(s)}{M(s)} - \frac{C(s)F(s)d(s)}{M(s)} \tag{F.214}$$

which is equal to:

$$u(s) = \frac{M(s)C_r(s)r(s) - M(s)\tilde{\sigma}(s) - C(s)F(s)d(s)}{M(s)(1 - C(s)) + C(s)F(s)} \tag{F.215}$$

From (F.177) it follows that:

$$u_{ref}(s) - u(s)$$
$$= \frac{M(s)C_r(s)r(s) - C(s)F(s)d_{ref}(s) - (M(s)C_r(s)r(s) - M(s)\tilde{\sigma}(s) - C(s)F(s)d(s))}{M(s)(1 - C(s)) + C(s)F(s)} \tag{F.216}$$

That is:

$$u_{ref}(s) - u(s) = \frac{M(s)\tilde{\sigma}(s) + C(s)F(s)\,(d(s) - d_{ref}(s))}{M(s)(1 - C(s)) + C(s)F(s)} \tag{F.217}$$

Equation (F.234) states that $\tilde{y}(s) = \frac{M(s)}{C(s)}\tilde{\sigma}(s)$, then:

$$u_{ref}(s) - u(s) = -H_2(s)(d_{ref}(s) - d(s)) + \frac{H_3(s)}{M(s)}\tilde{y}(s) \tag{F.218}$$

Above it can be found that $\|y_{ref} - y\|_{\mathcal{L}_\infty} < \gamma_1$. According to the Lipschitz conditions in Assumption F.1.4: $|d'(t, y_{ref}) - d'(t, y)| \leq L\,|(y_{ref} - y)|$. Then an upper bound can be found as:

$$\|(u_{ref} - u)_t\|_{\mathcal{L}_\infty} < \|H_2(s)\|_{\mathcal{L}_1} L\gamma_1 + \left\|\frac{H_3(s)}{M(s)}\right\|_{\mathcal{L}_1} \bar{\gamma}_0 = \gamma_2 \tag{F.219}$$

$\square$

F.5.10 Proof of Lemma F.5.1

By merging (F.213) into $y(s) = M(s)\,(u(s) + \sigma(s)) + y_{in}(s)$, one has:

$$y(s) = M(s)\,(C_r(s)r(s) - \tilde{\sigma}(s) + (1 - C(s))\sigma(s)) + y_{in}(s) \tag{F.220}$$

We have for $\sigma(s)$:

$$\sigma(s) = \frac{(F(s) - M(s))u(s) + F(s)d(s)}{M(s)} \tag{F.221}$$

$$\sigma(s) = \frac{(F(s) - M(s))(C_r(s)r(s) - \tilde{\sigma}(s) - C(s)\sigma(s)) + F(s)d(s)}{M(s)} \tag{F.222}$$

$$\sigma(s) = \frac{F(s) - M(s)}{M(s)}\,(C_r(s)r(s) - \tilde{\sigma}(s) - C(s)\sigma(s)) + \frac{F(s)}{M(s)}d(s) \tag{F.223}$$

$$\sigma(s) = \frac{(F(s) - M(s))(C_r(s)r(s) - \tilde{\sigma}(s)) + F(s)d(s)}{M(s)(1 - C(s)) + C(s)F(s)} \tag{F.224}$$

Substituting the last equation into (F.220):

$$\begin{aligned}
y(s) &= M(s)\left(C_r(s)r(s) - \tilde{\sigma}(s) + (1 - C(s))\frac{(F(s) - M(s))(C_r(s)r(s) - \tilde{\sigma}(s)) + F(s)d(s)}{M(s)(1 - C(s)) + C(s)F(s)} \right) \\
&\quad + y_{in}(s)
\end{aligned} \tag{F.225}$$

$$y(s) = H(s)C_r(s)r(s) - H(s)\tilde{\sigma}(s) + H(s)(1 - C(s))d(s) + y_{in}(s) \tag{F.226}$$

Equation (F.182) in section F.2 is defined as:

$$y_{ref}(s) = H(s)C_r(s)r(s) + H(s)(1 - C(s))d_{ref}(s) + y_{in}(s)$$

Consequently:

$$e(s) = y_{ref}(s) - y(s) = H(s)\tilde{\sigma}(s) + H(s)(1 - C(s))d_{err}(s) \tag{F.227}$$

where $d_{err}(s)$ is defined by $d_{ref}(s) - d(s)$.

Define temporarily $\eta_\sigma(s) = H(s)\tilde{\sigma}(s)$, then with $H(s)$ being strictly proper and stable and Assumption F.1.4 an upper bound can be derived as:

$$\|e_l\|_{\mathcal{L}_\infty} \leq \|H(s)(1 - C(s))\|_{\mathcal{L}_1} L \|e_l\|_{\mathcal{L}_\infty} + \|\eta_{\sigma,l}\|_{\mathcal{L}_\infty} \tag{F.228}$$

Lemma F.5.2.

$$H(s)\tilde{\sigma}(s) = H_2(s)\tilde{y}(s)$$

The proof can be found in section F.5.11.

Then:

$$\|e_l\|_{\mathcal{L}_\infty} \leq \|H(s)(1 - C(s))\|_{\mathcal{L}_1} L \|e_l\|_{\mathcal{L}_\infty} + \|H_2(s)\|_{\mathcal{L}_1} \|\tilde{y}_l\|_{\mathcal{L}_\infty} \tag{F.229}$$

which is:

$$\|e_l\|_{\mathcal{L}_\infty} \leq \frac{\|H_2(s)\|_{\mathcal{L}_1}}{1 - \|H(s)(1 - C(s))\|_{\mathcal{L}_1} L} \|\tilde{y}_l\|_{\mathcal{L}_\infty} \tag{F.230}$$

$\square$

F.5.11 Proof of Lemma F.5.2

The subtraction of

$$y(s) = M(s)(u(s) + \sigma(s)) + y_{in}(s) \tag{F.231}$$

and

$$\hat{y}(s) = M(s)u(s) + c^T (s\mathbb{I} - A)^{-1} \hat{\sigma}(s) + y_{in}(s) \tag{F.232}$$

results in:

$$\tilde{y}(s) = c^T (s\mathbb{I} - A)^{-1} \hat{\sigma}(s) - M(s)\sigma(s) \tag{F.233}$$

Reusing the definition of $\tilde{\sigma}(s)$ in Equation (F.211) of section F.5.10, namely:

$$\tilde{\sigma}(s) = \frac{C(s)}{M(s)} c^T (s\mathbb{I} - A)^{-1} \hat{\sigma}(s) - C(s)\sigma(s)$$

Gives:

$$\tilde{y}(s) = \frac{M(s)}{C(s)} \tilde{\sigma}(s) \tag{F.234}$$

Let this relation be rewritten to the form of:

$$H(s)\tilde{\sigma}(s) = \frac{C(s)}{C(s)} \frac{M(s)}{M(s)} H(s)\tilde{\sigma}(s) = H_2(s)\tilde{y}(s) \tag{F.235}$$

$\square$

Appendix G

Signal Hedging with Saturation

This chapter addresses the partition of the input signal, i.e. hedging of some signals contained in the input signal. Conditions are shown for a correct proportion concerning saturation. Saturation is an inherently nonlinear operation. The following is first presented in [2] by the author.

G.1 Introduction and Conditions

Define $sat_0(v)$ as one specific saturation operator of the signal v, with $l \leq v \leq u$, with l being its lower bound and u its upper bound. Every signal affected by this saturation operator is subject to the same upper and lower bound.

Let there be two signals a and b, whose saturated sum $c = sat_0(a + b)$ is entering system I as input. Both, signal a and signal b are individually available. Signal c is used as source of an input signal to system II but not directly sent to it. As *acceptable* is defined to hedge signal a in such a way from system II that both, systems I and II get the same fraction β of signal b, contained in the saturated sum c.

Theorem G.1.1.

Within this definition it is an *acceptable* solution to subtract signal a from the saturated sum *iff* a is not able to reach saturation alone at any time, i.e.:

$sat_0(a + b) - a$ is *acceptable, iff* $a = sat_0(a)$ $\forall a$.

Figure G.1 shows the situation.

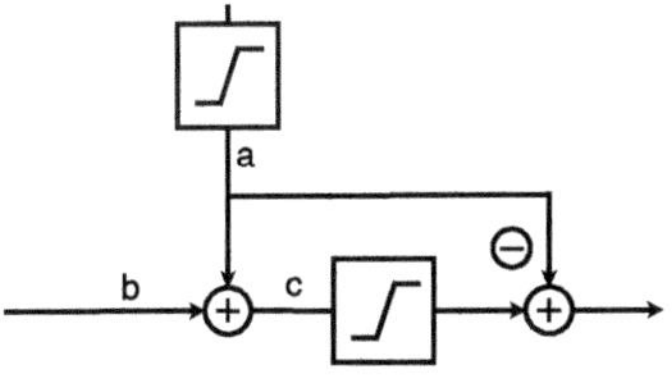

FIGURE G.1: Signal hedging with saturation

Note that a is not required to lie naturally within the saturation bounds, $a = sat_0(a)$ can be achieved artificially by simply applying the same saturation operator to a as to $a + b$ as shown in Figure G.1.

G.2   Proof of Theorem G.1.1

The following proof addresses combinations of:
1) Signal a lies within the saturation bounds or not.
2) Signal $a + b$ lies within the saturation bounds or not.
This results in four possible combinations.

Case 1: $sat_0(a) = a$ *AND* $sat(a + b) = a + b$
This case is trivial as no saturation is affecting any signal which ensures linearity of the operations: $\beta = b = sat_0(a + b) - a$

Case 2: $sat_0(a) = a$ *AND* $sat(a + b) \neq a + b$
Case 2.1: $b > 0 : \beta = u - a = sat_0(a + b) - a$, or $\beta + a = u$
Case 2.2: $b < 0 : \beta = l - a = sat_0(a + b) - a$, or $\beta + a = l$
both cases hold for $a > 0$ and $a < 0$, and satisfy the condition $l \leq sat_0(a + b) \leq u$. Figure G.2 illustrates this case. Note that for $a > 0$ the lower bound l can coincide with 0, but $a < 0$ is not meaningful for $l \geq 0$ as a would deceed the lower bound.

Case 3: $sat_0(a) \neq a$ *AND* $sat_0(a + b) = a + b$
Case 4: $sat_0(a) \neq a$ *AND* $sat_0(a + b) \neq a + b$
Cases 3 and 4: It is sufficient to show that these cases are not always *acceptable* by finding an example for at least one of them which violates the saturation condition. Thus, the upper saturation block (colored) in Figure G.1 is the one in question and virtually removed for a counterexample.

Counterexample: Let $a > u - l$ and $sat_0(a + b) = u$. Then apply the subtraction after the saturation operator, namely $sat(a + b) - a = u - a$ for a positive a, then with this and the inequality before it follows that $sat_0(a + b) < l$ which contradicts the saturation condition. If in the implementation $sat_0(a) \neq a$ is excluded, there is no chance to run into a not *acceptable* case.

This completes the proof of Theorem G.1.1.

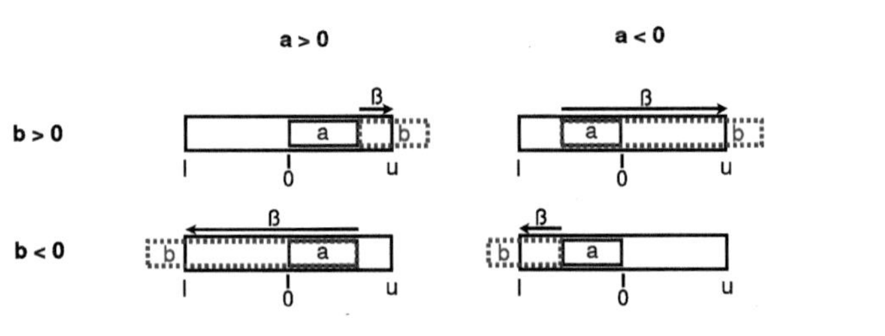

FIGURE G.2: Explanation to proof of Theorem G.1.1, Case 2

For the case of a standalone $\mathcal{L}_1$-controller, every signal except the adaptive control signal may be desired to be hedged from the predictor input. Then this structure sends as much of the adaptive signal to the predictor as is sent to the plant after saturation of the total control signal.

Appendix H

The Effect of a_{SP} on $|\tilde{x}|$, $|x_{ref} - x|$, $|x_{ref} - \hat{x}|$

This chapter examines the effects of the prediction error on performance and robustness in state feedback.

H.1 Description

In state feedback, the error dynamics are mainly shaped by a_{SP}. As long as it is stable and within reasonable numerical range, it can be chosen largely free. The impact on performance and robustness is evaluated next for scalar systems. For MIMO systems with unmatched uncertainties the statements made may no longer be true. This chapter refers to state feedback only as the degree of freedom a_{SP} does not exist in this form in output feedback.

H.2   Example

Consider the following system:

$$\text{Plant:} \quad \dot{x}(t) = -3x(t) + u(t) - 8 \tag{H.1}$$

$$\text{Control law:} \quad u(s) = -C(s)\hat{\sigma}(s) \tag{H.2}$$

$$\text{Low-pass filter:} \quad C(s) = \frac{15}{s + 15} \tag{H.3}$$

$$\text{Initial condition of low-pass filter:} \quad C_{in}(s) = 8 \tag{H.4}$$

$$\text{Sample time:} \quad T = 0.01s \tag{H.5}$$

$$\text{Command:} \quad r(t) = 0 \text{ for } t \in [0,2) \text{ and } r(t) = 1 \text{ for } t \in [2,7] \tag{H.6}$$

a_{SP} is defined differently in the two systems as -4 and -0.1, respectively:

$$\text{Predictor 1:} \quad \dot{\hat{x}}(t) = -3x(t) + u(t) + \hat{\sigma}(t) - 4\tilde{x}(t) \tag{H.7}$$

$$\text{Predictor 2:} \quad \dot{\hat{x}}(t) = -3x(t) + u(t) + \hat{\sigma}(t) - 0.1\tilde{x}(t) \tag{H.8}$$

The adaptive laws are then defined as:

$\hat{\sigma}(t) = \frac{-4e^{-4 \cdot 0.01}}{(1 - e^{-4 \cdot 0.01})} \tilde{x}(t)$ and $\hat{\sigma}(t) = \frac{-0.1e^{-0.1 \cdot 0.01}}{(1 - e^{-0.1 \cdot 0.01})} \tilde{x}(t)$, respectively.

A gain $= 3$ is applied to the command to achieve steady state tracking.

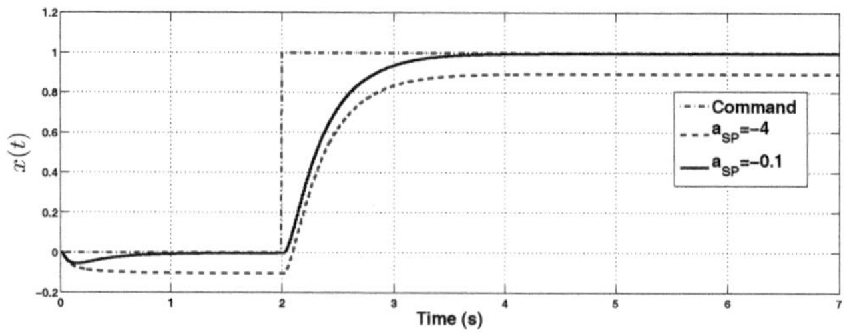

FIGURE H.1: Sensitivity of a_{SP} on $|r(t) - x(t)|$

It can be seen from Figures H.1, H.2, and H.3 that $\sigma(t)$ has significant impact on the trajectories of $x(t)$, $\hat{x}(t)$, $\tilde{x}(t)$. Disturbances cause a deviation of $x(t)$ from the command. The faster a_{SP}, the larger this offset. At the same time, slower a_{SP} cause $\hat{x}(t)$ to move away from

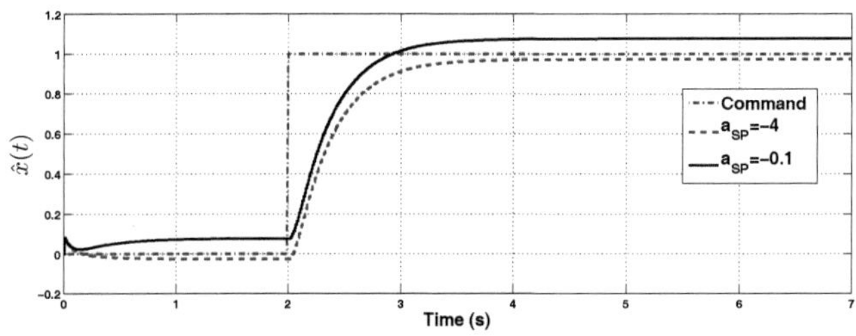

FIGURE H.2: Sensitivity of a_{SP} on $|r(t) - \hat{x}(t)|$

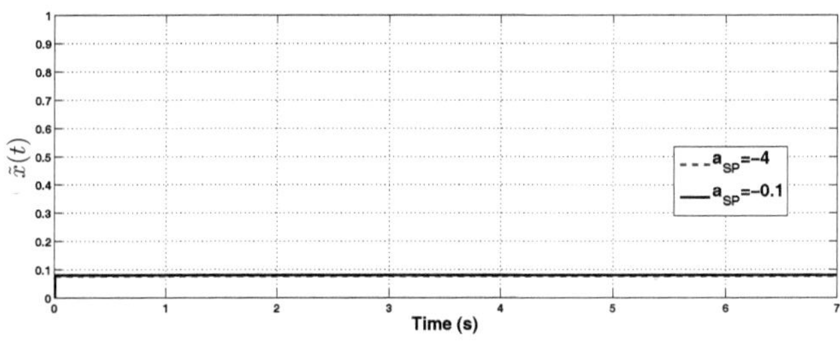

FIGURE H.3: Sensitivity of a_{SP} on $|\tilde{x}(t)|$

the command. The differences $x_{(a_{SP}=-4)}(t) - x_{(a_{SP}=-0.1)}(t)$ and $\hat{x}_{(a_{SP}=-4)}(t) - \hat{x}_{(a_{SP}=-0.1)}(t)$ are roughly the same. In other words, a change in a_{SP} causes only small changes in $\tilde{x}(t)$.

The initialization of the low-pass filter is only for avoiding an initial negative peak of the trajectory of $x(t)$. By hiding the inertness of the low-pass filter, a better insight in the trajectory deviation is obtained.

These statements show that slow a_{SP} have performance benefits. The same simulation is used to obtain the time delay margin dependent on a_{SP}. To this end, a time delay is added to the plant input, and only there. This means the input of the predictor has a time offset relative to the input of the plant.

Figure H.4 shows that the time delay margin is for both systems $\tau \in [100ms, 110ms]$. The faster a_{SP}, the higher the time delay margin.

This shows the trade-off between performance and robustness. For this example, the time delay margin with faster a_{SP} is increased by estimated $(5...10)\%$, accompanied by a significantly increased tracking error $|r(t) - x(t)|$, in particular in $\|r(t) - x(t)\|_{\mathcal{L}_2}$.

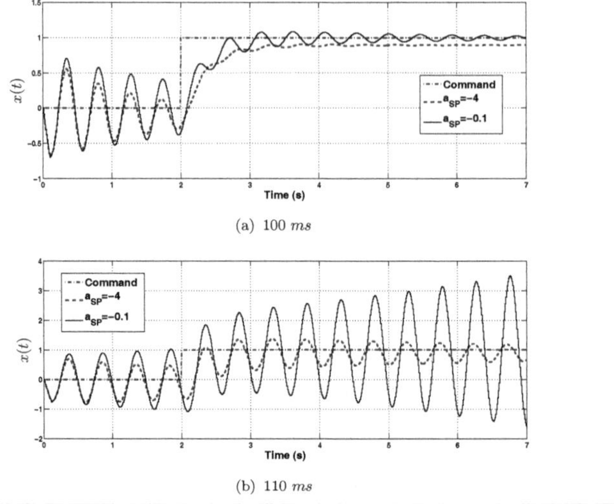

(a) 100 *ms*

(b) 110 *ms*

FIGURE H.4: Effect of a_{SP} on robustness with additional delay

H.3   Explanation

In general we have:

$$\text{Plant:} \quad \dot{x}(t) = ax(t) + b(u(t) + \sigma(t)) \tag{H.9}$$

$$\text{Predictor:} \quad \dot{\hat{x}}(t) = ax(t) + b(u(t) + \hat{\sigma}(t)) + a_{SP}\tilde{x}(t) \tag{H.10}$$

$$\text{Error dynamics:} \quad \dot{\tilde{x}}(t) = a_{SP}\tilde{x}(t) + b\hat{\sigma}(t) - b\sigma(t) \tag{H.11}$$

Integrating the error dynamics in the span $t \in [0, T]$ we have:

$$\tilde{x}(iT + t) = e^{a_{SP}t}\tilde{x}(iT) + \int_0^t e^{a_{SP}(t-\tau)}b\hat{\sigma}(iT)d\tau - \int_0^t e^{a_{SP}(t-\tau)}b\sigma(iT + \tau)d\tau \tag{H.12}$$

Use (2.20) as adaptive law – a gain that is a constant term for this integral. Then, in Equation (H.12) only the first two terms cancel by applying $\hat{\sigma}(iT)$ from the adaptive law. The other term – dependent on a_{SP} – is assumed to be bounded but in general nonzero after one step.

Let the remaining term that is not canceled down in Equation (H.12) be denoted as R, then:

$$R := \tilde{x}((i_0 + 1)T) = - \int_0^T e^{a_{SP}(T-\tau)} b\sigma(i_0 T + \tau) d\tau \tag{H.13}$$

For the simulation example of this chapter, integrating (H.13) gives $R = 0.0784$ for $a_{SP} = -4$ and $R = 0.08$ for $a_{SP} = -0.1$, cf. Figure H.3.

For a better insight, see Figure H.5. It shows the process in a continuous system applying the piece-wise constant adaptive law. The data for this figure coincide with the simulation above except for a_{SP} where the case of $a_{SP} = -4$ is changed to $a_{SP} = -40$ for readability. If it was kept $a_{SP} = -4$, the lines in the first subplot would lie on top of each other in this figure size. The line for $a_{SP} = -40$ in the third term thus becomes a little smaller, confer numbers from above.

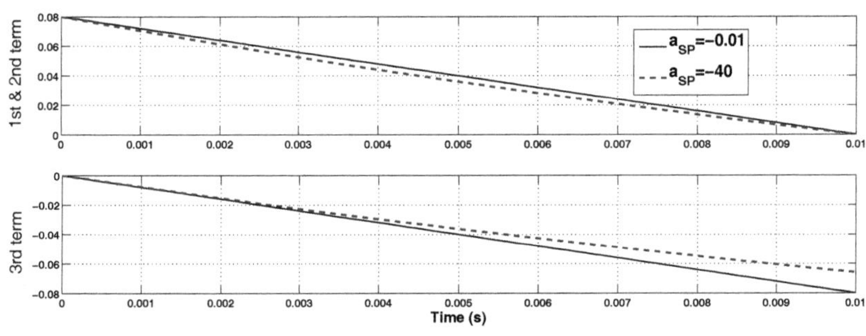

FIGURE H.5: Effects of a_{SP} on various terms in the error dynamics

The first plot shows the sum of the first two terms from the right hand side in Equation (H.12), and the second one the third term. Note that both diagram parts show exponential functions despite $\sigma(t)$ is assumed to be constant. Both trajectories start at the same value $\tilde{x}(iT) = \tilde{x}_0$ and the sum of the first to terms of both reaches zero at the same time due to the inversion of the error dynamics.

If it can be shown that a_{SP} in some region has very small sensitivity on $|\tilde{x}(t)|$, then it can be shown generally that higher values of $|a_{SP}|$ cause a higher offset of $x(t)$ from $x_{ref}(t)$, and a limit of $|x_{ref}(t) - \hat{x}(t)|$ is proportional to $|\tilde{x}(t)|$. This is done next:

Sensitivity of a_{SP} on $|\tilde{x}(t)|$:
As the remaining term R leads directly to $\tilde{x}(t)$, the goal is to find the sensitivity of a_{SP} on the remaining term R.

It is assumed that $\sigma(t)$ is constant over one T for a sensitivity analysis. It may change however between different iT – a discrete approximation of a continuous function with the sampling time step T. This is an legitimate assumption as changes of $\sigma(t)$ over one T are assumed to be small.

Solving Equation (H.13) for a constant $\sigma(\tau)$ gives:

$$R = \frac{\sigma}{a_{SP}} \left(1 - e^{a_{SP}T}\right) \tag{H.14}$$

The partial derivative of R to a_{SP} is:

$$\frac{\partial R}{\partial a_{SP}} = -\frac{\sigma}{a_{SP}^2} \left(e^{a_{SP}T}(a_{SP}T - 1) + 1\right) \tag{H.15}$$

Then we have:

$$\lim_{T \to 0} \frac{\partial R}{\partial a_{SP}} = 0 \tag{H.16}$$

This is shown in Figure H.6.

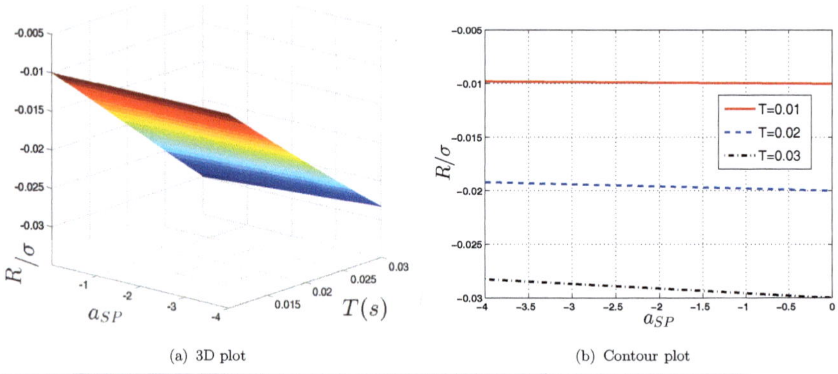

(a) 3D plot   (b) Contour plot

FIGURE H.6: Sensitivity a_{SP} to $\tilde{x}(t)$

In Numbers:

T	0.01	0.02	0.03
Change	2%	4%	6%

where "Change" [%] denotes: $\dfrac{R/\sigma(a_{SP}=-4) - R/\sigma(a_{SP}=-0.01)}{R/\sigma(a_{SP}=-4)}$

In words: The smaller T, the smaller this gradient. This explains the very small differences of $\tilde{x}(t)$ in Figure H.3 for different a_{SP}.

Deviation $|x_{ref}(t) - x(t)|$

This relation is described in Equation (E.68), Appendix E.4.3.

For convenience it is repeated here:

$$\|(x_{ref} - x)_i\|_{\mathcal{L}_\infty} \leq \frac{\|(s-a)^{-1}bC(s)(s-a_{SP})/b\|_{\mathcal{L}_1}}{1 - \|(s-a)^{-1}b(1-C(s))\|_{\mathcal{L}_1} L} \|\tilde{x}_i\|_{\mathcal{L}_\infty}$$

With $\tilde{x}(t)$ approximately not being influenced by a_{SP}, the influence of a_{SP} to $|x_{ref}(t) - x(t)|$ lies in $(s - a_{SP})$ such that a_{SP} determines directly $|x_{ref}(t) - x(t)|$, note the DC-gain. The faster a_{SP}, the bigger the difference $|x_{ref}(t) - x(t)|$.

Deviation $|x_{ref}(t) - \hat{x}(t)|$

$$x_{ref}(t) - \hat{x}(t) = x_{ref}(t) + x(t) - x(t) - \hat{x}(t) = x_{ref}(t) - x(t) - \tilde{x}(t) \tag{H.17}$$

Which explains that the difference between $x_{ref}(t) - \hat{x}(t)$ and $x_{ref}(t) - x(t)$ is $\tilde{x}(t)$.

Note that $x_{ref}(t)$ features a virtual ideal adaptive law, hence it is without steady state error to $r(t)$.

$$\lim_{T\to\infty, a_{SP}\to 0} (x_{ref}(t) - x(t)) = 0 \quad \to \quad \lim_{T\to\infty, a_{SP}\to 0} (x_{ref}(t) - \hat{x}(t)) = -\tilde{x}(t) \tag{H.18}$$

where of course $\lim_{T\to\infty} \tilde{x}(t) = 0$, too. It means that the closer a_{SP} to zero, the smaller $|x_{ref}(t) - x(t)|$ and the closer $x(t)$ to $r(t)$.

> In conclusion, slow a_{SP} are better performing, but less robust in terms of the time delay margin.

This is valid for scalar systems. In MIMO systems an additional DOF adds expressed by imaginary parts. In the adaptation loop, a low damped pair of poles exists due to the high-gain adaptive law. Then, a k_{sp} can add damping, see Appendix D for the terminology. A systematic design like LQR (linear-quadratic regulator) can be applied to move the lightly damped pole pairs to some velocity and damping value. Damping the adaptation loop slows down adaptation with some loss of information. More robustness and a smoother control signal however can be expected besides influences on measurement noise, at least for some specific adaptive laws.

A proof why slower a_{SP} are hurting the time delay margin is not shown here due to the lack of a general proof for it in nonlinear, unknown systems.

In output feedback a_{SP} does not exist in this form. The effects of propagating disturbances are analog. For this reason, a modified, recursive adaptive law is applied, assuring $\tilde{y}(t)$ is kept small despite large (constant) disturbances.

Appendix I

Controller Robustness against Time Step Variations

This chapter evaluates to what extent deviations from the nominal hardware time step length present performance or robustness issues. It refers to state feedback with the raw adaptive law. The aspects shown can be extended to the recursive adaptive law and output feedback – similar effects occur.

I.1   Description

The computational time step length T is hard-coded in the adaptive law. It is impossible in a physical CPU to meet this time step exactly. After one T the first two terms of Equation (2.19) are canceled. By deviating from the nominal step length, an additional error is introduced as cancellation of the first two terms is no longer operating correctly. This error is similar to the one appearing in a quantized system.

I.2   Example

For Figure I.1 and Figure I.2, the following system is used:

$$\text{Plant:} \quad \dot{x}(t) = -1x(t) + 0.8u(t) + 10 \tag{I.1}$$

$$\text{Predictor:} \quad \dot{\hat{x}}(t) = -8x(t) + 0.33(u(t) + \hat{\sigma}(t)) + a_{SP}\tilde{x}(t) \tag{I.2}$$

$$\text{Adaptive law:} \quad \hat{\sigma}(t) = (1/0.33)a_{SP}e^{a_{SP}T}/(1 - e^{a_{SP}T})\tilde{x}(t) \tag{I.3}$$

$$\text{Low-pass filter:} \quad C(s) = \frac{15}{s + 15} \tag{I.4}$$

The following anomalies are introduced:

	Implemented time step T [s]	**Simulated time step T_{CPU} [s]**
System 1 (Figure I.1)	0.01	0.025
System 2 (Figure I.2)	0.01	0.005

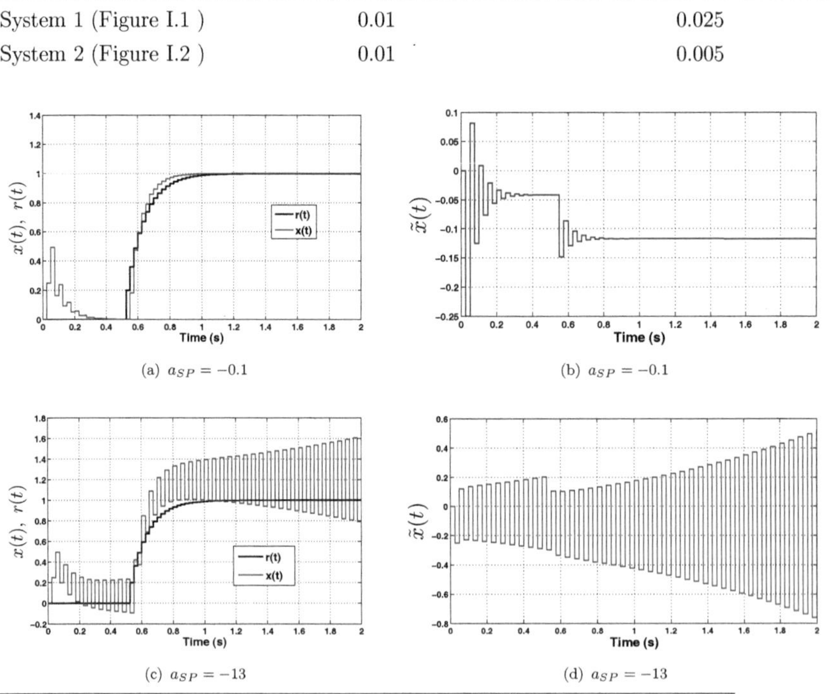

(a) $a_{SP} = -0.1$

(b) $a_{SP} = -0.1$

(c) $a_{SP} = -13$

(d) $a_{SP} = -13$

FIGURE I.1: Time step variation: CPU step length (0.025), slower than hard-coded step length (0.01).

It can be seen that a smaller a_{SP} is more robust against a variation in the time step of the hardware. Note that the time step in the code is less than half and double, respectively of what the CPU conducts. The critical case here is a slower running CPU than what

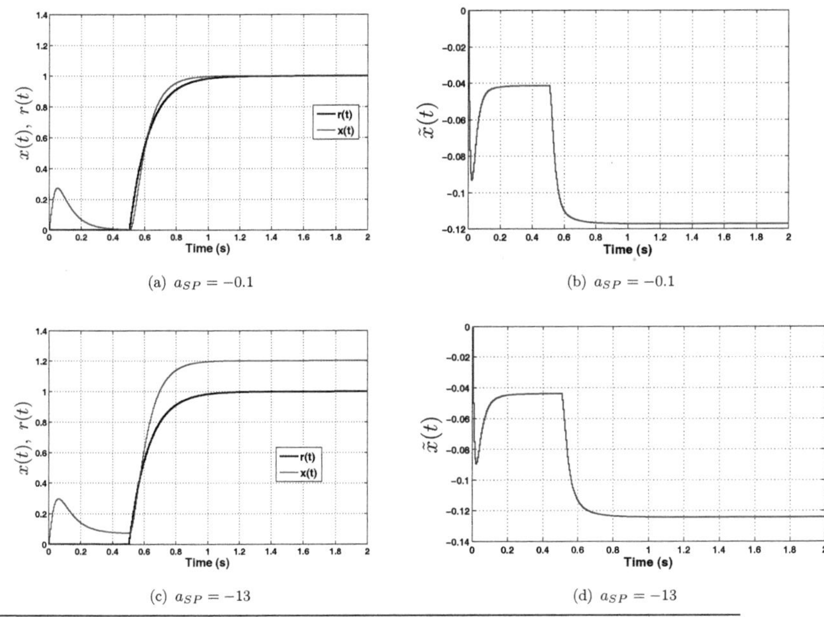

(a) $a_{SP} = -0.1$

(b) $a_{SP} = -0.1$

(c) $a_{SP} = -13$

(d) $a_{SP} = -13$

FIGURE I.2: Time step variation: CPU step length (0.005), faster than hard-coded step length (0.01).

is assumed. Such extreme variations however are not expected. Discretization errors of dynamic filters are significant at such an error level.

I.3   Explanation

The term $\chi(t = T_{CPU})$ in Equation (E.30) is:

$$\chi((i+1)T_{CPU}) = e^{a_{SP}T_{CPU}}\tilde{x}(iT_{CPU}) + \frac{b}{-a_{SP}}(1 - e^{a_{SP}T_{CPU}})\hat{\sigma}(iT_{CPU})$$

Where $\hat{\sigma}(iT_{CPU})$ is:

$$\hat{\sigma}(iT_{CPU}) = \frac{a_{SP}e^{a_{SP}T}}{b(1 - e^{a_{SP}T})}\tilde{x}(iT_{CPU}) \tag{I.5}$$

These equations together are not zero anymore but:

$$\chi((i+1)T_{CPU}) = e^{a_{SP}T_{CPU}}\tilde{x}(iT) + \frac{b(1 - e^{a_{SP}T_{CPU}})}{-a_{SP}}\frac{a_{SP}}{b(1 - e^{a_{SP}T})}e^{a_{SP}T}\tilde{x}(iT_{CPU}) \qquad (I.6)$$

That is:

$$\chi((i+1)T_{CPU}) = \left(e^{a_{SP}T_{CPU}} - \frac{(1 - e^{a_{SP}T_{CPU}})}{(1 - e^{a_{SP}T})}e^{a_{SP}T} \right)\tilde{x}(iT_{CPU}) \qquad (I.7)$$

This adds a very small additional constant to $\chi((i+1)T_{CPU})$ and widens the performance bounds, but does not lead immediately to instability unless the controller is too close at its robustness limit.

If the total execution time can be measured, a scheduled adaptive law adjusting to different T is imaginable.

The same analysis can be conducted for output feedback, for which circumstances are very similar to what is shown in this chapter.

Appendix J

Controller Robustness against Measurement Noise

This section addresses the measurement noise propagation to the control signal and the system states or the raw output (without new noise). The sensor noise includes measurement inherent noise as well as signal corruption by measuring structural effects or vibrations.
A continuous approximation in the form of continuous transfer functions is provided. These are not valid for high frequency noise but approximate the effects.
It is shown that state and output feedback as introduced in this thesis are similar in their reactions to measurement noise.

J.1 Description

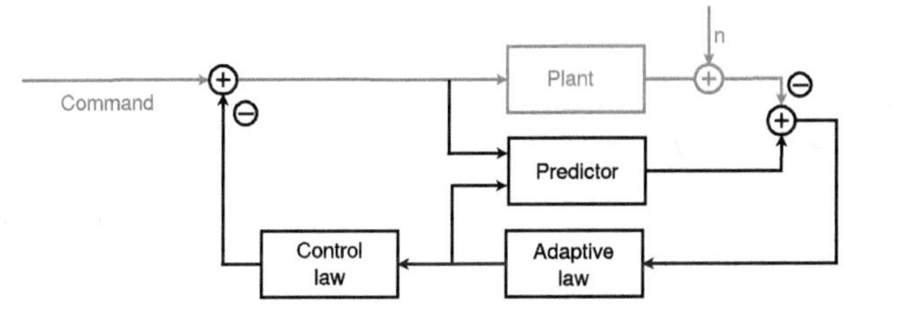

FIGURE J.1: Closed loop $\mathcal{L}_1$-controller with noise

While disturbances act directly on $\dot{x}(t)$, sensor noise is added to $x(t)$ and the raw $y(t)$, respectively. Figure J.1 defines how measurement is introduced into the loop.

State Feedback:

Let the following scalar system be given:

$$\text{Plant:} \quad \dot{x}(t) = ax(t) + bu(t) \tag{J.1}$$

$$\text{Desired dynamics:} \quad M(s) := \frac{b}{s-a} \tag{J.2}$$

$$\text{Predictor:} \quad \dot{\hat{x}}(t) = ax(t) + b(u(t) + \hat{\sigma}(t)) + a_{SP}\tilde{x}(t) \tag{J.3}$$

$$\text{Control law:} \quad u(s) = -C(s)\hat{\sigma}(s) \tag{J.4}$$

$$\text{Prediction error:} \quad \tilde{x}(s) = \hat{x}(s) - x(s) - \nu(s) \tag{J.5}$$

$$\text{Adaptive gains:} \quad F_1 := \frac{a_{SP}e^{a_{SP}T}}{b(1 - e^{a_{SP}T})}, \quad F_2 := \frac{a_{SP}}{T \cdot b(1 - e^{a_{SP}T})}, \tag{J.6}$$

$$F(s) := F_1 + \frac{F_2}{s} \tag{J.7}$$

$$\text{Raw adaptive law:} \quad \hat{\sigma}(t) = F_1\tilde{x}(t) \tag{J.8}$$

$$\text{Recursive adaptive law:} \quad \hat{\sigma}(t) = F(t)\tilde{x}(t) \tag{J.9}$$

Remark J.1.1.

The definition $F(s)$ in (J.7) is derived from the fact that $h(iT) = -\tilde{x}(iT) + h((i-1)T)$ and $T \cdot (-\tilde{x}(iT) + h((i-1)T)) = -\int \tilde{x}(t)$, i.e. $h(t)T$ is a discrete time integration of $-\tilde{x}(t)$. To the raw adaptive law only F_1 applies. The recursive law adds a term with F_2 to the total $F(s)$.

Theorem J.1.1.

With the raw adaptive law, the effects of noise $\nu(t)$ to the states $x(t)$ and input $u(t)$ – approximated by continuous transfer functions – in state feedback are:

$$x(s) = \frac{bF_1C(s)}{s - bF_1 - a_{SP} + a}\nu(s); \quad u(s) = \frac{F_1C(s)(s-a)}{s - bF_1 - a_{SP} + a}\nu(s)$$

The proof can be found in section J.3 .

Theorem J.1.2.

With the recursive adaptive law, the effects of noise $\nu(t)$ to the states $x(t)$ and input $u(t)$ – approximated by continuous transfer functions – in state feedback are:

$$x(s) = \frac{b\left(F_1 + \frac{F_2}{s}\right)C(s)}{s - b\left(F_1 + \frac{F_2}{s}\right) - a_{SP} + a}\nu(s); \quad u(s) = \frac{\left(F_1 + \frac{F_2}{s}\right)C(s)(s-a)}{s - b\left(F_1 + \frac{F_2}{s}\right) - a_{SP} + a}\nu(s)$$

The proof can be found in section J.4 .

Output Feedback:

$$\text{Plant:} \quad y(s) = M(s)u(s) \tag{J.10}$$

$$\text{Predictor:} \quad \hat{y}(s) = M(s)u(s) + M_{um}(s)\hat{\sigma}(s) \tag{J.11}$$

$$\text{with:} \quad M(s) = c^T(sI - A)^{-1}b, \quad M_{um}(s) = c^T(sI - A)^{-1} \tag{J.12}$$

$$\text{Control law:} \quad u(s) = -C(s)\frac{M_{um}(s)}{M(s)}\hat{\sigma}(s) =: -L(s)\hat{\sigma}(s) \tag{J.13}$$

$$\text{Adaptive gains:} \quad F_1 := -\Phi^{-1}(T)\Upsilon(T)1_1, \quad F_2 := -\Phi^{-1}(T)\frac{1_1}{T} \tag{J.14}$$

$$F(s) := F_1 + \frac{F_2}{s} \tag{J.15}$$

$$\text{Recursive adaptive law:} \quad \hat{\sigma}(t) = F(t)\tilde{y}(t) \tag{J.16}$$

where $\Phi(T) = -(\Lambda A\Lambda^{-1})^{-1}\left(\mathbb{I} - e^{\Lambda A\Lambda^{-1}T}\right)\Lambda$, $\Upsilon(T) = e^{\Lambda A\Lambda^{-1}T}$, and $1_1 = [1, 0, ..., 0]^T \in \mathbb{R}^n$.

Theorem J.1.3.

The effects of noise $\nu(t)$ to the output $y(t)$ and input $u(t)$ – approximated by continuous transfer functions – are:

$$y(s) = \frac{M(s)L(s)\left(F_1 + \frac{F_2}{s}\right)}{1 - M_{um}(s)\left(F_1 + \frac{F_2}{s}\right)}\nu(s), \quad u(s) = \frac{L(s)\left(F_1 + \frac{F_2}{s}\right)}{1 - M_{um}(s)\left(F_1 + \frac{F_2}{s}\right)}\nu(s)$$

In [43] similar results are shown for state feedback, however for different adaptive laws and without the recursive adaptive law. In this case the denominator shows a lightly damped pole pair which is responsible for a peak in this transfer function from measurement noise to the state. This noise peak does not exist for the piece-wise constant adaptive law as the denominator does not present the lightly damped pole pair.

J.2   Example

State Feedback:

The following example for state feedback is used:

$$\text{Plant:} \quad \dot{x}(t) = -3x(t) + 3u(t) \tag{J.17}$$

$$\text{Predictor:} \quad \dot{\hat{x}}(t) = -3x(t) + 3(u(t) + \hat{\sigma}(t)) + a_{SP}\tilde{x}(t) \tag{J.18}$$

$$\text{Sample time:} \quad T = 0.01s \tag{J.19}$$

The variations under consideration are:

$$\text{Low-pass filter 1:} \quad C_1(s) = \frac{15}{s+15} \tag{J.20}$$

$$\text{Low-pass filter 2:} \quad C_2(s) = \left(\frac{23}{s+23}\right)^2 \tag{J.21}$$

$$a_{SP}\ 1: \quad a_{SP,1} = -4 \tag{J.22}$$

$$a_{SP}\ 2: \quad a_{SP,2} = -0.1 \tag{J.23}$$

The second low-pass filter is chosen to have approximately the same bandwidth as the first filter. The transfer functions from noise to the control signal and to the system states are shown in Figure J.2 and Figure J.3 for all combinations.

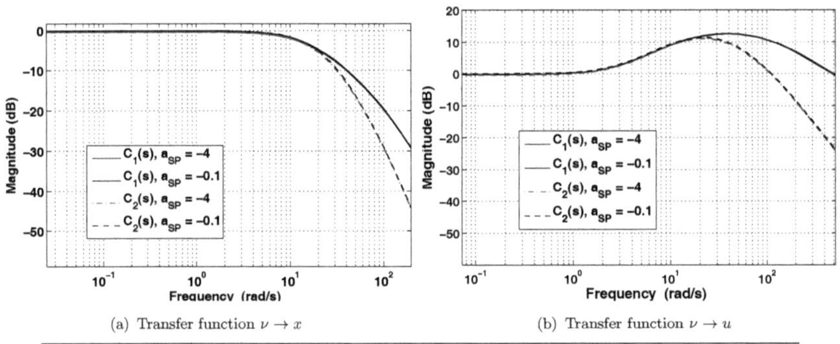

(a) Transfer function $\nu \to x$ (b) Transfer function $\nu \to u$

FIGURE J.2: Transfer functions of noise in state feedback with raw adaptive law

The low-pass filter is the decisive element. Variations in a_{SP} hardly make any difference.

Output Feedback:

Figure J.4 uses the following parameters:

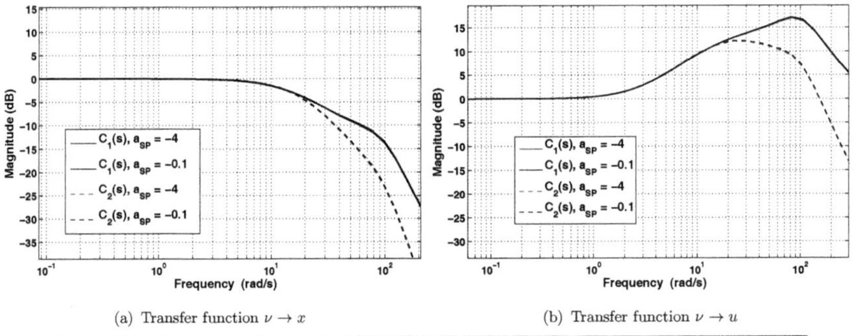

(a) Transfer function $\nu \to x$ (b) Transfer function $\nu \to u$

FIGURE J.3: Transfer functions of noise in state feedback with recursive adaptive law

$$\text{Desired Dynamics} \quad M(s) = \frac{7.5(s+20)}{(s+10)(s+15)} \tag{J.24}$$

$$\text{Sample time:} \quad T = 0.01 \ s \tag{J.25}$$

The variations under consideration are:

$$\text{Low-pass filter 1:} \quad C_1(s) = \frac{15}{s+15} \tag{J.26}$$

$$\text{Low-pass filter 2:} \quad C_2(s) = \left(\frac{23}{s+23}\right)^2 \tag{J.27}$$

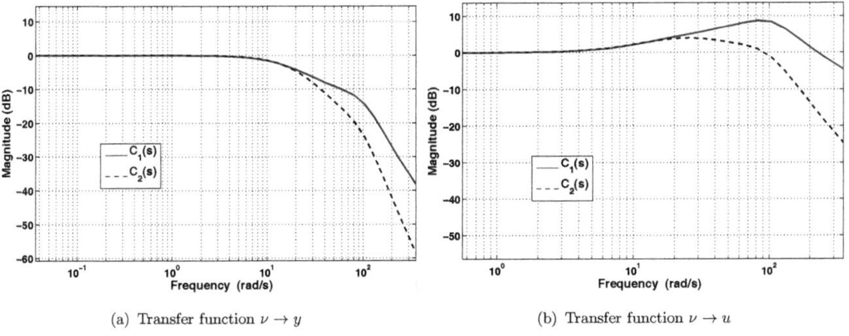

(a) Transfer function $\nu \to y$ (b) Transfer function $\nu \to u$

FIGURE J.4: Transfer functions of noise in output feedback

J.3 Proof of Theorem J.1.1

This proof applies to scalar systems which are affected by measurement noise $\nu(t)$. Neither a command nor uncertainties are included as those do not change the reaction to noise.

The predictor is:

$$\dot{\hat{x}}(t) = a(x(t) + \nu(t)) + b(u(t) + \hat{\sigma}(t)) + a_{SP}\tilde{x}(t) \tag{J.28}$$

Rewriting the predictor with $x(t) = \hat{x}(t) - \nu(t) - \tilde{x}(t)$ yields:

$$\dot{\hat{x}}(t) = a\hat{x}(t) - a\tilde{x}(t) + b(u(t) + \hat{\sigma}(t)) + a_{SP}\tilde{x}(t) \tag{J.29}$$

In Laplace domain, this can be written as:

$$\hat{x}(s) = \frac{bu(s) + b\hat{\sigma}(s) + (a_{SP} - a)\tilde{x}(s)}{s - a} \tag{J.30}$$

The plant is in Laplace domain:

$$x(s) = \frac{b}{s - a}u(s) \tag{J.31}$$

With the last two equations, the error dynamics with $\tilde{x}(s) = \hat{x}(s) - x(s) - \nu(s)$ are:

$$\tilde{x}(s) = \frac{b\hat{\sigma}(s) + (a_{SP} - a)\tilde{x}(s)}{s - a} - \nu(s) \tag{J.32}$$

By noting that $\hat{\sigma}(s) = F_1\tilde{x}(s)$, we have:

$$\tilde{x}(s) = -\frac{\nu(s)}{1 - \frac{bF_1 + a_{SP} - a}{s - a}} \tag{J.33}$$

That is:

$$\tilde{x}(s) = -\frac{s - a}{s - bF_1 - a_{SP} + a}\nu(s) =: H_{\tilde{x}\nu,raw}(s)\nu(s) \tag{J.34}$$

Using this result in $u(s) = -C(s)F_1\tilde{x}(s)$ and $x(s) = -M(s)C(s)F_1\tilde{x}(s)$ delivers:

$$x(s) = \frac{M(s)C(s)F_1(s - a)}{s - bF_1 - a_{SP} + a}\nu(s) = \frac{bF_1C(s)}{s - bF_1 - a_{SP} + a}\nu(s) \tag{J.35}$$

and:

$$u(s) = \frac{F_1 C(s)(s-a)}{s - bF_1 - a_{SP} + a} \nu(s) \tag{J.36}$$

$\square$

J.4 Proof of Theorem J.1.2

By substituting F_1 by $F(s)$ in (J.34), (J.35), and (J.36), Theorem J.1.2 follows directly.

$\square$

J.5 Proof of Theorem J.1.3

This proof applies to SISO transfer functions which are affected by measurement noise $\nu(t)$. Neither a command nor uncertainties are included as those do not change the reaction to noise.

Note that $L(s)$ is a row vector of transfer functions and $F(s)$ is a column vector of gains. Thus, multiplication is *not* commutative.

The error dynamics are defined as:

$$\tilde{y}(s) = \hat{y}(s) - y(s) - \nu(s) \tag{J.37}$$

Taking the equations above into account, this is:

$$\tilde{y}(s) = -M(s)L(s)F(s)\tilde{y}(s) + M_{um}(s)F(s)\tilde{y}(s) + M(s)L(s)F(s)\tilde{y}(s) - \nu(s) \tag{J.38}$$

This can be simplified to be:

$$\tilde{y}(s) = \frac{-\nu(s)}{1 - M_{um}(s)F(s)} =: H_{\tilde{y}\nu,rec}(s)\nu(s) \tag{J.39}$$

Using this result in $y(s) = -M(s)L(s)F(s)\tilde{y}(s)$ and $u(s) = -L(s)F(s)\tilde{y}(s)$, we have with $F(s) = F_1 + \frac{F_2}{s}$:

$$y(s) = \frac{M(s)L(s)\left(F_1 + \frac{F_2}{s}\right)}{1 - M_{um}(s)\left(F_1 + \frac{F_2}{s}\right)} \nu(s) \tag{J.40}$$

and:

$$u(s) = \frac{L(s)\left(F_1 + \frac{F_2}{s}\right)}{1 - M_{um}(s)\left(F_1 + \frac{F_2}{s}\right)}\nu(s) \tag{J.41}$$

$\square$

The transfer function $L(s)$ has the same relative degree as $C(s)$. This can be seen as $M(s)$ is $M_{um}(s)$ multiplied by a constant vector b.

For Appendix F, the following transfer functions are derived:

From the predictor equation (F.54) we have without uncertainties/disturbances:

$$\tilde{x}(s) = (s\mathbb{I} - A)^{-1}\hat{\sigma}(s) \tag{J.42}$$

Furthermore, $\hat{\sigma}(s) = F(s)\tilde{y}(s)$.

Substituting Equation (J.39) for $\tilde{y}(s)$ delivers:

$$\nu_{\tilde{x}}(s) = \frac{-(s\mathbb{I} - A)^{-1}F(s)}{1 - M_{um}(s)F(s)}\nu(s) =: H_{\tilde{x}\nu}(s)\nu(s) \tag{J.43}$$

Appendix K

Modeling of Inertia Effects with a Gyroscopic Term

This section describes a method to add more knowledge about the physical system to the control structures. One term considering inertia effects of a rigid body for rotational dynamics and another modeling the rotor as a gyroscope are shown. A helicopter rotor is clearly not a gyroscope, but the paradigm presented here implies that modeling a rotor as gyroscope is closer to reality than handling the rotor as if it did not spin. For the sake of simplicity, the system rotor-fuselage is modeled as two connected bodies, where the only relative movement between them is the main rotation of the rotor in relation to the fuselage, i.e. tilting the rotor disk against the fuselage is not considered. The following definitions apply:

I_H: Inertia tensor of the helicopter without rotor;

I_R: Inertia tensor of the rotor;

V_B: Angular momentum denoted in frame B;

ω^{BR}: Angular velocity of frame R with respect to frame B;

ω^{IB}: Angular velocity of frame B with respect to frame I;

n_R : Number of rotor blades;

As the movement of the rotor approximately around the rotor mast, namely the z-component of $(\omega^{BR})_B$, is the dominant one, it is sufficient here to neglect all terms but I_{zz} ("moment of inertia around the mast") in I_R, the inertia tensor of the main rotor.

The angular momentum denoted in the body-fixed frame is:

$$(V)_B = (I_H + I_R)(\omega^{IB})_B + I_R(\omega^{BR})_B \tag{K.1}$$

with $(\omega^{BR})_B = \begin{bmatrix} 0 & 0 & \omega_r \end{bmatrix}^T_B$

Newton's second law is only valid with respect to an inertial frame. Then the derivative of $(V)_B$ to frame I is:

$$
\begin{aligned}
(M^I)_B &= \left(\frac{d}{dt}\right)^I (V)_B \\
&= (\dot{V})^B_B + (\omega^{IB})_B \times (V)_B \\
&= (I_H + I_R)(\dot{\omega}^{IB})^B_B + (\omega^{IB})_B \times \left[(I_H + I_R)(\omega^{IB})_B + I_R(\omega^{BR})_B\right] \\
&= 0
\end{aligned}
\tag{K.2}
$$

It follows that:

$$
(\dot{\omega}^{IB})^B_B = -(I_H + I_R)^{-1}\left[(\omega^{IB})_B \times \left[(I_H + I_R)(\omega^{IB})_B + I_R(\omega^{BR})_B)\right]\right] =: U(t)
\tag{K.3}
$$

With the term $I_R(\omega^{BR})_B$ being optional as gyroscopic term. The inertia of the rotor can be sized up with the mass m and length l of each rotor blade with a Steiner term to $I_R = n_R\left(ml^2/12 + ml^2/4\right) = n_R ml^2/3$ (inertia for a infinitely thin rod)

Then an expanded state predictor would be:

$$
\dot{\omega}(t) = A\omega(t) + U(t) + B(u(t) + \hat{\sigma}(t)) + A_{SP}\tilde{\omega}(t)
\tag{K.4}
$$

Note that this predictor is inevitably a MIMO system. A possible baseline controller for decoupling this predictor would be $u_{BSL,d}(t) = -B^{-1}U(t)$.

Appendix L

Alternative Predictor in an Error Space

This section shows the shift from controlling the plant to controlling the error between the plant and the desired system. The new "plant" to be controlled is therefore the tracking error dynamics. A state or an output predictor containing the desired tracking error dynamics replaces the standard predictor. Figure L.1 shows a possible structure.

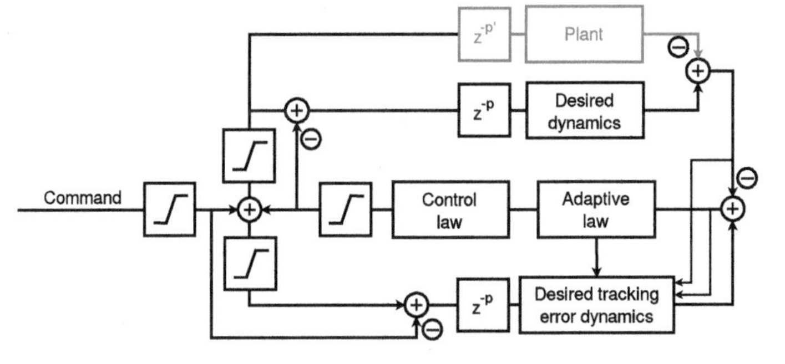

FIGURE L.1: Controller structure for error space

Multiple saturation operators are necessary to enable the subtraction of the adaptive control signal and the command, respectively. See Appendix G for a detailed explanation.

The desired *tracking* error dynamics are defined as the difference between desired dynamics and a plant model (or something close to it) – provided that the desired dynamics are different from the as stable identified nominal plant. Tuning of the desired tracking error dynamics largely excludes its DC-gain, as it is determined by the desired dynamics and the plant model. Instead, the structure of the desired dynamics as well as its bandwidth, DC-gain, and order constitute the tuning parameters. The desired tracking error dynamics however have to be close to the deviation of the desired dynamics from the plant model as this states the natural and thus feasible dynamics, provided that the desired dynamics are robustly feasible themselves. The implementation of explicit command dynamics is necessary (in contrast to a command alone). The output of this command model is subtracted from the system output to properly catch the tracking error dynamics. Otherwise the control goal is unrealistic which leads to poor performance and robustness. In general, faster tracking error dynamics lead to higher performance but less robustness against time delays.

The motivation of this structure lies in defining the tracking error dynamics independently of the desired dynamics. In addition, instead of accounting for non-minimum phase behavior of the plant in the input channel by e.g. additional time delay, the non-minimum phase behavior can be taken explicitly into account in this architecture by rendering the desired dynamics as well as the assumed dynamics non-minimum phase; the actual non-minimum phase behavior is then hidden by the error.

However, this structure has several shortcomings:
As explained above, the degree of freedom of choosing the error dynamics independently of the desired dynamics is very limited. Additionally, this method adds a great deal of complexity. The tuning effort rises as defining desired tracking error dynamics is much less intuitive than defining the desired plant dynamics. The implementation effort also rises significantly. Furthermore, the deviation of the assumed plant dynamics from the desired dynamics have to be nonzero and reasonable, adding another cumbersome constraint.

Appendix M

Alternative State Predictor Enclosing Baseline Controller States

This section refers to an alternative state predictor for an extended state space in augmentation. The new full state space embraces states of the aircraft and the baseline controller. These additional states emerge from integrals (integral of attitude errors) and lead-lag filters at the baseline controller output. The baseline controller renders the closed-loop system higher order than the open-loop system.

M.1 Definitions

The new augmented system matrix is denoted as A_a.

The new state vector is $x(t) = \left[\omega^T(t), z^T(t) \right]^T$.

Where:

$$x(t) = \begin{bmatrix} \omega(t) \\ z(t) \end{bmatrix} \in \mathbb{R}^n, \quad \text{with} \quad \omega(t) = \begin{bmatrix} p(t) \\ q(t) \\ r(t) \end{bmatrix} \in \mathbb{R}^3, \quad z(t) = \begin{bmatrix} z_1(t) \\ z_2(t) \\ z_3(t) \end{bmatrix} \in \mathbb{R}^{n-3}$$

Note that $z_k(t)$ are vectors, $k \in \{1, 2, 3\}$.

The baseline controller is fictively separated into two groups:
1) The feedback loop of angular rates;
2) The feedforward term for angular rates together with the entire attitude loop (including the integrator of the attitude error);

Then, part "2)" of this separation acts as a command for the plant, to which the feedback loop of part "1)" is applied. This is the same idea as in a cascaded controller and is equivalent to the baseline architecture shown.

Define a pseudo command (i.e. part "2)"): $\check{r}(s) = ATT(s) + (LL_{FFW}(s) \cdot K_{d,cmd} \cdot \omega_{cmd}(s))$

where $ATT(s) \in \mathbb{R}^3$ is the signal vector from all attitude terms in the baseline controller, i.e a filtered PI formulation of the attitude error, $LL_{FFW}(s) \in \mathbb{R}^{3\times3}$ denotes a diagonal matrix of feedforward filters for the angular rates, $K_{d,cmd} \in \mathbb{R}^{3\times3}$ is the diagonal matrix of feedforward gains for angular rates and $\omega_{cmd}(s) \in \mathbb{R}^3$ are the commanded angular rates.

Further define G_h^k as the product of all filters at the output of the baseline controller for each axis $k \in \{pitch, roll, yaw\}$. Notch filters are not considered part of G_h^k as they are tailored to the final processing of the total actuator command vector rather than loop shaping of the feedback controller loops.

The state space formulation of the baseline controller for each axis involves the matrices A_k, B_k, C_k, and D_k, with $k \in \{pitch, roll, yaw\}$.

With this, define:

$$A_{TF} = \begin{bmatrix} A_1 & 0 & 0 \\ 0 & A_2 & 0 \\ 0 & 0 & A_3 \end{bmatrix}; \quad B_{TF} = \begin{bmatrix} B_1 & 0 & 0 \\ 0 & B_2 & 0 \\ 0 & 0 & B_3 \end{bmatrix}; \quad C_{TF} = \begin{bmatrix} C_1 & 0 & 0 \\ 0 & C_2 & 0 \\ 0 & 0 & C_3 \end{bmatrix};$$

$$D_{TF} = \begin{bmatrix} D_1 & 0 & 0 \\ 0 & D_2 & 0 \\ 0 & 0 & D_3 \end{bmatrix}; \quad K_{d\omega} = \begin{bmatrix} k_{dq} & 0 & 0 \\ 0 & k_{dp} & 0 \\ 0 & 0 & k_{dr} \end{bmatrix};$$

Further define:

$$A_a = \begin{bmatrix} A_{(3\times3)} & 0_{(3\times n-3)} \\ -B_{TF}K_{d\omega} & A_{TF} \end{bmatrix}; \quad B_r = \begin{bmatrix} 0_{(3\times3)} \\ B_{TF} \end{bmatrix}; \quad B_\sigma = \begin{bmatrix} B_{(3\times3)} & 0_{(3\times n-3)} \\ 0_{(n-3\times3)} & B_{um} \end{bmatrix};$$

$$B_u = \begin{bmatrix} B_{(3\times3)} \\ 0_{(n-3\times3)} \end{bmatrix};$$

M.2   Core Statements

To introduce the dynamics of the baseline controller into the predictor, the baseline controller is formulated in a state space representation for the three axes pitch, roll, yaw, where the pseudo command $\check{r}$ is defined above (dependencies (s) and (t) omitted for simplicity):

Axis	Baseline controller is a SISO transfer function in each axis:	Baseline controller in state space representation:
Pitch	$u^1_{BSL} = G^1_h(\check{r}_1 - k_{dq}q_{comp})$	$\begin{aligned} \dot{z}_1 &= A_1 z_1 + B_1(\check{r}_1 - k_{dq}q_{comp}) \\ u_{BSL,1} &= C_1 z_1 + D_1(\check{r}_1 - k_{dq}q_{comp}) \end{aligned}$
Roll	$u^2_{BSL} = G^2_h(\check{r}_2 - k_{dp}p_{comp})$	$\begin{aligned} \dot{z}_2 &= A_2 z_2 + B_2(\check{r}_2 - k_{dp}p_{comp}) \\ u_{BSL,2} &= C_2 z_2 + D_2(\check{r}_2 - k_{dp}p_{comp}) \end{aligned}$
Yaw	$u^3_{BSL} = G^3_h(\check{r}_3 - k_{dr}r_{comp})$	$\begin{aligned} \dot{z}_3 &= A_3 z_3 + B_3(\check{r}_3 - k_{dr}r_{comp}) \\ u_{BSL,3} &= C_3 z_3 + D_3(\check{r}_3 - k_{dr}r_{comp}) \end{aligned}$

Then by lumping all dimensions:

$$\dot{z}(t) = A_{TF}z(t) + B_{TF}(\check{r}(t) - K_{d\omega}\omega_{comp}(t)) \tag{M.1}$$

$$u_{BSL}(t) = C_{TF}z(t) + D_{TF}(\check{r}(t) - K_{d\omega}\omega_{comp}(t)) \tag{M.2}$$

Merging the state space representation into the general state predictor formulation, one has:

$$\text{Plant:} \quad \dot{x}(t) = A_a x(t) + B_r \check{r}(t) + B_\sigma \sigma(t) + B_u u(t) \tag{M.3}$$

$$\text{Predictor:} \quad \dot{\hat{x}}(t) = A_a x(t) + B_r \check{r}(t) + B_\sigma \hat{\sigma}(t) + B_u u(t) + A_{SP}\tilde{x}(t) \tag{M.4}$$

$$\text{Error dynamics:} \quad \dot{\tilde{x}}(t) = A_{SP}\tilde{x}(t) + B_\sigma \tilde{\sigma}(t) \tag{M.5}$$

$$\text{Input:} \quad u(t) = u_{BSL}(t) + u_{\mathcal{L}_1}(t), \quad u_{\mathcal{L}_1}(t) = -C(s)\hat{\sigma}_{(1:3\times1)}(t) \tag{M.6}$$

$$\text{Adaptive law:} \quad \hat{\sigma}(iT) = -B_\sigma^{-1}(A_{SP}^{-1}(e^{A_{sp}T_s} - \mathbb{I}_n))^{-1}e^{A_{sp}T_s}\tilde{x}(iT) =: F\tilde{x}(iT) \tag{M.7}$$

The state predictor can thus be written as:

$$\begin{bmatrix} \dot{\omega}(t) \\ \dot{z}(t) \end{bmatrix} = \begin{bmatrix} A_{(3\times3)} & 0_{(3\times n-3)} \\ -B_{TF}K_{d\omega} & A_{TF} \end{bmatrix} \begin{bmatrix} \omega(t) \\ z(t) \end{bmatrix} + \begin{bmatrix} 0_{(3\times3)} \\ B_{TF} \end{bmatrix} [\check{r}(t)]$$
$$+ \begin{bmatrix} B_{(3\times3)} & 0_{(3\times n-3)} \\ 0_{(n-3\times3)} & B_{um} \end{bmatrix} [\hat{\sigma}(t)] + \begin{bmatrix} B_{(3\times3)} \\ 0_{(n-3\times3)} \end{bmatrix} [u(t)] + [A_{SP}] \begin{bmatrix} \tilde{\omega}(t) \\ \tilde{z}(t) \end{bmatrix} \tag{M.8}$$

where B_{um} can be chosen as e.g. $B_{um} = \frac{tr[B]}{3} \cdot \mathbb{I}_{(n-3 \times n-3)}$. Its choice is rather arbitrary, but as B_{um} is part of B_σ, whose inversion is computed in the adaptive law, numerically well suited values are to be chosen. The choice above ensures that B_{um} has the same magnitude as the average diagonal entry of B.

The vector $\tilde{x}(t)$ is gained by integrating the left-hand side of Equation (M.8) and its comparison to the measured $\omega(t)$ and $z(t)$ from the baseline controller, i.e. the couplings (non-diagonal entries) in the predictor are responsible for a $\tilde{z}(t)$ to be existent.

Choose A_{SP} e.g. as the desired dynamics, resulting from the baseline controller applied to a plant model:

$$A_{SP} = A_a + \begin{bmatrix} B_{(3 \times 3)} \\ 0_{(n-3 \times 3)} \end{bmatrix} \begin{bmatrix} -D_{TF} K_{d\omega} & C_{TF} \end{bmatrix} \tag{M.9}$$

Parameter Scheduling:

To allow for a balanced comparison of the non-extended state predictor with the extended one in Equation (M.8), parameter scheduling is transferred to the extended one. That means that A and B in Equation (M.8) are scheduled, where scheduling of A_{SP} can be done but for saving computational power as well be omitted since the difference observed can be very small compared to a scheduled A_{SP}. The adaptive law however needs to be scheduled with the predictor in this case to avoid introducing errors into the loop. Note that an online calculation of the adaptive law is to be avoided in a real application due to a required matrix inversion in the adaptive law.

M.3 Necessity of the Alternative Predictor

Theorem M.3.1. Both, the extended and the simple state predictor can be reduced to the form

$$\dot{\hat{\omega}}(t) = A\omega(t) + \Xi(t)\tilde{\omega}(t)$$

where only $\Xi(t)$ is different for the two forms.

The proof can be found in section M.4.

Theorem M.3.1 suggests that with the choice of the respective A_{SP}, the simple (non-extended) state predictor can approximate the extended one. Although a relatively weak statement, it can often be verified in simulations.

Moreover, several shortcomings arise with the extended predictor:

Depending on number and structure of filters in the baseline controller, the dimension of the extended state predictor matrices can become large. With the baseline controller described in this thesis, the dimension would be at least $n = 14$ compared to $n = 3$ for a simple state predictor not including the baseline controller states.

Scheduling of A and B is much more difficult to realize. This adds massively to computational effort.

Decoupling functions are difficult to include if realized as dynamic systems. If decoupling functions are implemented as transfer functions (which is the case in the baseline controller at hand), they would add additional states into the scheme. Again, this would increase the computational effort.

M.4 Proof of Theorem M.3.1

It follows from Equation (M.8) that:

$$\dot{\hat{\omega}}(t) = A\omega + B\left(\hat{\sigma}_1(t) + u(t)\right) + A_{SP,11}\tilde{\omega}(t) + A_{SP,12}\tilde{z}(t) \tag{M.10}$$

$$\dot{\hat{z}}(t) = \underbrace{-B_{TF}K_{d\omega}\omega(t) + A_{TF}z(t) + B_{TF}\check{r}(t)}_{\dot{z}(t)} + B_{um}\hat{\sigma}_2(t) + A_{SP,21}\tilde{\omega}(t) + A_{SP,22}\tilde{z}(t) \tag{M.11}$$

Thus:

$$\dot{\tilde{z}}(t) = B_{um}\hat{\sigma}_2(t) + A_{SP,21}\tilde{\omega}(t) + A_{SP,22}\tilde{z}(t) \tag{M.12}$$

Merging the Laplace transformed Equation (M.12) into Equation (M.10) implies:

$$\dot{\hat{\omega}}(t) = A\omega(t) + B\left(\hat{\sigma}_1(t) + u(t)\right)$$
$$+ A_{SP,11}\tilde{\omega}(t) + A_{SP,12}\left(\mathscr{L}^{-1}\left\{(sI - A_{SP,22})^{-1}\right\}(B_{um}\hat{\sigma}_2(t)) + A_{SP,21}\tilde{\omega}(t)\right)$$

(M.13)

$$\dot{\hat{\omega}}(t) = A\omega(t) + B\left(\hat{\sigma}_1(t) + u(t)\right)$$
$$+ \left(A_{SP,11} + A_{SP,12}\mathscr{L}^{-1}\left\{(sI - A_{SP,22})^{-1}\right\}A_{SP,21}\right)\tilde{\omega}(t)$$
$$+ A_{SP,12}\mathscr{L}^{-1}(sI - A_{SP,22})^{-1}B_{um}\hat{\sigma}_2(t)$$

(M.14)

where $\mathscr{L}^{-1}\{\cdot\}$ is the inverse Laplace transformation and $\hat{\sigma}_2(t)$ is a piece-wise constant function of $\tilde{\omega}(t)$:

$$\begin{bmatrix} \hat{\sigma}_1(iT) \\ \hat{\sigma}_2(iT) \end{bmatrix} = \begin{bmatrix} F_{11,(3\times3)} & F_{12,(3\times(n-3))} \\ F_{21,((n-3)\times3)} & F_{22,((n-3)\times(n-3))} \end{bmatrix} \begin{bmatrix} \tilde{\omega}(iT) \\ \tilde{z}(iT) \end{bmatrix}$$

(M.15)

Hence:

$$\hat{\sigma}_1(iT) = F_{11}\tilde{\omega}(iT) + F_{12}\tilde{z}(iT)$$

(M.16)

and:

$$\hat{\sigma}_2(iT) = F_{21}\tilde{\omega}(iT) + F_{22}\tilde{z}(iT)$$

(M.17)

For rewriting Equation (M.14), two elements are needed:

1) $\tilde{z}(s)$ **with** $\hat{\sigma}_2(s)$:

$$\tilde{z}(s) = (sI - A_{SP,22})^{-1}\left(B_{um}F_{21}\tilde{\omega}(s) + B_{um}F_{22}\tilde{z}(s) + A_{SP,21}\tilde{\omega}(s)\right)$$

(M.18)

2) $u(s)$:

$$u(s) + \hat{\sigma}_1(s) = (1 - C(s))\hat{\sigma}_1(s) = (1 - C(s))(F_{11}\tilde{\omega}(s) + F_{12}\tilde{z}(s))$$

(M.19)

$$\tilde{z}(s) = \left(I - (sI - A_{SP,22})^{-1}B_{um}F_{22}\right)^{-1}(sI - A_{SP,22})^{-1}\left(B_{um}(s)F_{21} + A_{SP,21}\right)\tilde{\omega}(s) \quad \text{(M.20)}$$

$$\tilde{z}(s) = P(s)\tilde{\omega}(s)$$

(M.21)

$$P(s) = \left(1 - (sI - A_{SP,22})^{-1}B_{um}F_{22}\right)^{-1}(sI - A_{SP,22})^{-1}(B_{um}F_{21} + A_{SP,21}) \tag{M.22}$$

Then, (M.14) is:

$$s\hat{\omega}(s) = A\omega(s) + [B(1 - C(s))(F_{11} + F_{12}P(s)) + A_{SP,11} + A_{SP,12}(sI - A_{SP,22})^{-1}A_{SP,21} \\ + A_{SP,12}(sI - A_{SP,22})^{-1}B_{um}(F_{21} + F_{22}P(s))]\tilde{\omega}(s) \tag{M.23}$$

For the simple predictor we have:

$$s\hat{\omega}(s) = A\omega(s) + (B(1 - C(s))F_0 + A_{SP})\tilde{\omega}(s) \tag{M.24}$$

where F_0 is the adaptive law in the non-extended predictor, such that $\hat{\sigma}(iT) = F_0\tilde{\omega}(iT)$.

Comparing Equation (M.23) and Equation (M.24) completes the proof of Lemma M.3.1.

$\square$

M.5 Example

The following system is used for a simple simulation, shown in Figure M.1 and Figure M.2:

$$
\begin{aligned}
\text{Plant:} \quad & \dot{\omega}(t) = \omega(t) + \omega^2(t) + 6u(t) \\
\text{Sampling time:} \quad & T = 0.01 \ s \\
\text{Command:} \quad & r(t) = 0 \text{ for } t \in [0, 1) \text{ and } r(t) = 1 \text{ for } t \in [1, 7] \\
\text{Low-pass filter:} \quad & C(s) = \tfrac{10}{s+10} \\
\text{Extended state definition:} \quad & \dot{z}(t) = r(t) - \omega(t) \\
\text{Baseline controller:} \quad & u_{BSL}(t) = K_P\dot{z}(t) + K_I z(t)
\end{aligned}
$$

System 1, simple predictor:

$$\text{Predictor: } \dot{\hat{\omega}}(t) = a\omega(t) + b(u(t) + \hat{\sigma}(t)) + a_{SP}\tilde{\omega}(t) \tag{M.25}$$

$$\text{Adaptive law: } \hat{\sigma}(iT) = \frac{a_{SP}e^{a_{SP}T}}{b(1 - e^{a_{SP}T})}\tilde{\omega}(iT) \tag{M.26}$$

$$\text{Control law: } u(s) = -C(s)\hat{\sigma}(s) + u_{BSL}(s) \tag{M.27}$$

System 2, extended predictor:

$$\text{Predictor:} \quad \begin{bmatrix} \dot{\hat{\omega}}(t) \\ \dot{\hat{z}}(t) \end{bmatrix} = \begin{bmatrix} a & 0 \\ -1 & 0 \end{bmatrix} \begin{bmatrix} \omega(t) \\ z(t) \end{bmatrix} + \begin{bmatrix} 0 \\ 1 \end{bmatrix} r(t) \tag{M.28}$$

$$+ \begin{bmatrix} b \\ 0 \end{bmatrix} u(t) + \begin{bmatrix} 1 & 0 \\ 0 & 1 \end{bmatrix} \hat{\sigma}(t) + A_{SP} \begin{bmatrix} \tilde{\omega}(t) \\ \tilde{z}(t) \end{bmatrix} \tag{M.29}$$

$$\text{Control law:} \quad u(s) = -C(s)\hat{\sigma}_1(s) + u_{BSL}(s) \tag{M.30}$$

where $\hat{\sigma}_1(s)$ is the first component of $\hat{\sigma}(s) \in \mathbb{R}^{2\times 1}$. Substituting $u_{BSL}(s)$ in (M.29) leads to the desired system matrix, which defines A_{SP}:

$$A_{SP} = \begin{bmatrix} a - bk_P & k_I \\ -1 & 0 \end{bmatrix} \tag{M.31}$$

The following parameters are used in the simulations: $K_P = K_I = 3$, $a = -1$, $b = 1$.

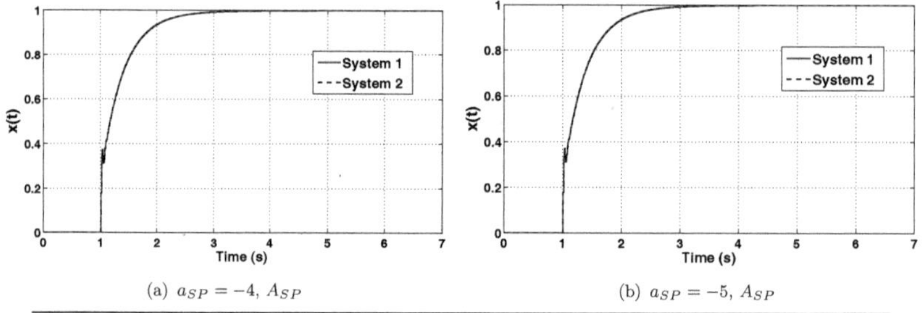

(a) $a_{SP} = -4$, A_{SP} (b) $a_{SP} = -5$, A_{SP}

FIGURE M.1: Trajectory $x(t)$ in augmentation, additional time delay of 0 ms

This simulation shows 1) that systems exist where an equivalent a_{SP} in the simple predictor can be found that resembles the behavior of the extended predictor, and 2) that faster a_{SP} are less performing but more robust. Any additional time delay is added to the input of the plant without modeling delay for the predictor. The relatively low time delay margin is explained by the huge deviation of the plant from the desired dynamics.

For comparison, a standalone $\mathcal{L}_1$-controller with $a = -4$ resembling the desired dynamics from above with $a_{SP} = -4$ and $a_{SP} = -0.4$ is shown in Figure M.3.

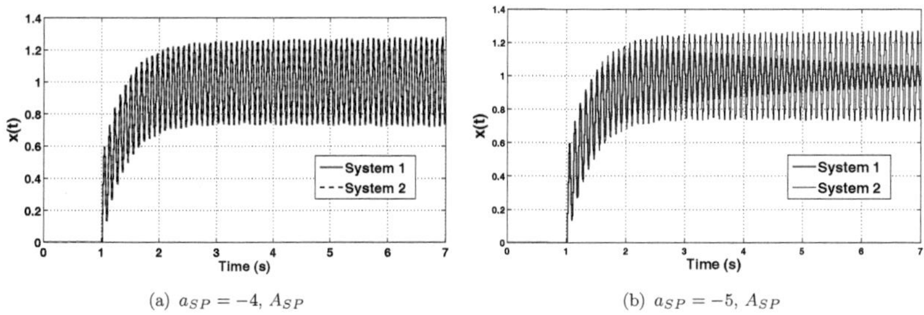

(a) $a_{SP} = -4$, A_{SP} (b) $a_{SP} = -5$, A_{SP}

FIGURE M.2: Trajectory $x(t)$ in augmentation, additional time delay of 10 ms

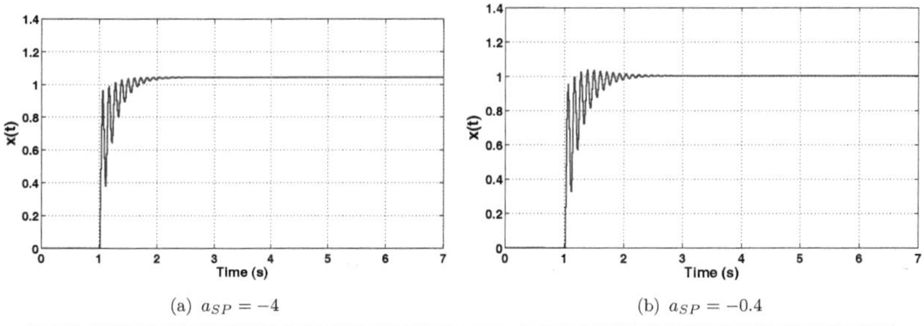

(a) $a_{SP} = -4$ (b) $a_{SP} = -0.4$

FIGURE M.3: Trajectory $x(t)$ in standalone mode, additional time delay of 10 ms

Appendix N

Examples of Practical Verification and Falsification

This chapter shows examples of verification – or falsification where applicable. Terminology used refers to state feedback, but the facts hold for output feedback, too.

Verification of a Computed $\tilde{x}(t)$ with Simulations

The verification for the piece-wise constant law and $\tilde{x}(t)$ is implicitly done in Appendix H, where the computed remaining term $R = 0.0784$ in Equations (H.13) and (H.14) coincides with the simulated one in Figure H.3. This example is not suitable to be transferred to the vertical speed controller for instance, as it presumes the absence of any other uncertainties – which is not true by many reasons for the vertical speed controller, starting with the higher order of the plant. Only a mass change affects directly the open-loop acceleration of the helicopter, which corresponds to a constant disturbance.

Verification of the Magnitude of $\tilde{x}(t)$

Since $\tilde{x}(t)$ behaves as predicted, the magnitude of $\tilde{x}(t)$ is to be checked, relative to $x(t)$. The size of $|\tilde{x}(t)|$ amounts inversely to the adaptation quality and overall controller performance. This verifies the implementation as opposed to the validation of the concept, e.g. whether desired handling qualities are achieved.

The following example of Figure N.1 is a manually piloted flight. The maneuver is a deceleration from 50 kts to 5 kts with a temporary pitch angle of 35 deg, $T = 1/(50\ Hz)$.

Verification of Predictor Dynamics with Identified Plant

The next example, shown in Figure N.2, refers to the choice of the first order predictor dynamics fitted to the identified plant for augmentation – or if a smooth version of the nominal dynamics is the desired dynamics in a standalone controller.

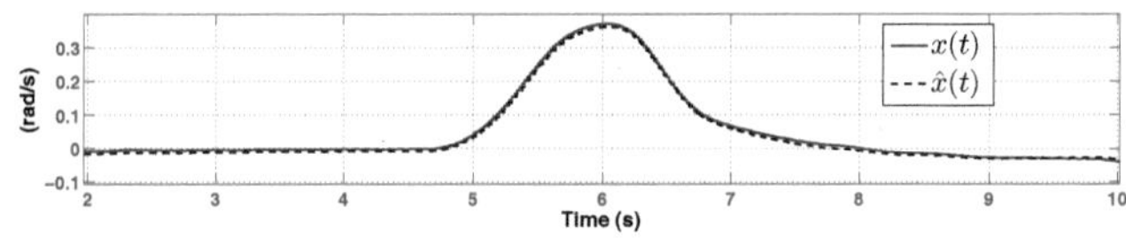

FIGURE N.1: Exemplary time histories of $x(t)$ and $\hat{x}(t)$ for pitch rate

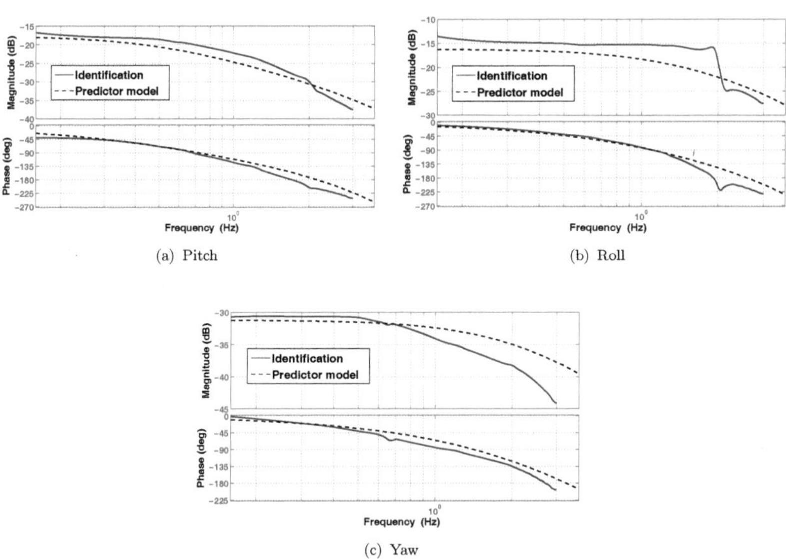

FIGURE N.2: Comparison of predictor dynamics to identified plant

Thus, the matching of the predictor with the identified response is verified or falsified –
depending on the requirements. The identified frequency response itself can be verified
with the plant, e.g. in time domain.

Compatibility of the Baseline Controller to the Predictor Dynamics

Figure N.3 reviews compatibility of the baseline controller to the predictor dynamics. This
is of importance as the baseline controller is not designed upon an explicit model, especially
not one used as predictor. The bode diagrams in Figure N.3 are produced with system
identification of a closed loop system of the baseline controller and the predictor (wherein
the predictor plays the role of the plant). The alternative to gain these bode diagrams
would be to analytically merge the transfer functions of the baseline controller and of the
predictor. However, with the system identification, actuator rate and position saturations

of the actuators are taken into account. Only decoupling functions from the baseline controller are removed.

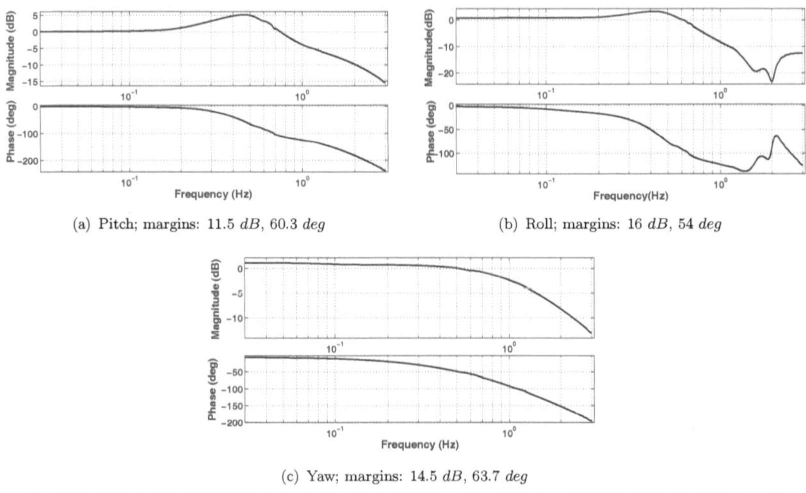

(a) Pitch; margins: 11.5 *dB*, 60.3 *deg*

(b) Roll; margins: 16 *dB*, 54 *deg*

(c) Yaw; margins: 14.5 *dB*, 63.7 *deg*

FIGURE N.3: Applying the baseline controller to predictor dynamics – bode diagrams of command and output

In pitch and roll, an overshoot is introduced; it is questionable whether a first order predictor is a good choice for augmentation of the baseline controller shown. This is especially true as the baseline controller is not designed directly for the model used in the predictor.

Appendix O

Simulation Setup

This chapter delineates the simulation. The objective of the implementation is to provide a test environment and demonstrator. Affinity to real hardware is sought e.g. by a discrete implementation.

O.1   Concept

The simulation is composed of the **simulation kernel**, scripts and code in various **programming languages and environments**, and a **visualization** module.

The **simulation kernel** is a proprietary, gray-box, precompiled simulation of the helicopter dynamics. Its input is the actuator output, its output are the flight conditions like angular rates, attitudes, altitude, speed etc.. Parameters such as helicopter mass, CG, air density, cable length to and mass of external loads, and others can be defined. Neither equations nor any data e.g. in form of look-up tables are known about the simulator. It is treated like a helicopter prototype whose dynamics are not known yet, but can be evaluated with e.g. system identification. It is known however that the simulation can be considered high-fidelity. It comprehends certain blade modes and an engine model. An actuator model is not included.

The simulation kernel is integrated in a graphical and plain-text based **development environment** which the controller is developed in and interfaces to input devices like a low-cost joystick and to a visualization, namely "FlightGear" are implemented.

Connected via UDP (user datagram protocol), FlightGear is used for **visualization** only, its simulation engine is idle. Helicopter type (for visualization) and environmental parameters are defined in a script. For enhancing the visualization, the FlightGear simulation is run with activated head-up display. Furthermore, a picture-in-picture mode is implemented to display an exterior view in the lower right corner. Figure O.1 shows two screenshots of the visualization.

(a) Cockpit view in picture-in-picture mode

(b) Exterior view

FIGURE O.1: Visualization screenshots

O.2 Structure

The simulation unifies the following options:
If not stated otherwise, items refer to pitch, roll, yaw.

- Turn coordination

- Rate command

- Attitude command

- Velocity command mode for hover

- Baseline controller

- $\mathcal{L}_1$-controller

    - Outer loop for standalone $\mathcal{L}_1$

    - State feedback with raw adaptive law

    - State feedback with recursive adaptive law

    - State feedback including the baseline controller states

    - State feedback in "error space"

    - Output feedback with recursive adaptive law

- $\mathcal{L}_1$ state feedback for vertical speed with raw adaptive law

- $\mathcal{L}_1$ state feedback for vertical speed with recursive adaptive law

- Aerobatics mode (deactivates attitude control and turn coordination)

- Input signal before or after actuators

- Slung loads

- Interpolation or analytic function for parameter scheduling

- Adding an inertia term plus optionally an gyroscopic term

- Sensor noise simulation

- Engine failure simulation

- Selectable time step length for controller greater than $0.001\ s$, independent of sampling time in the simulation kernel

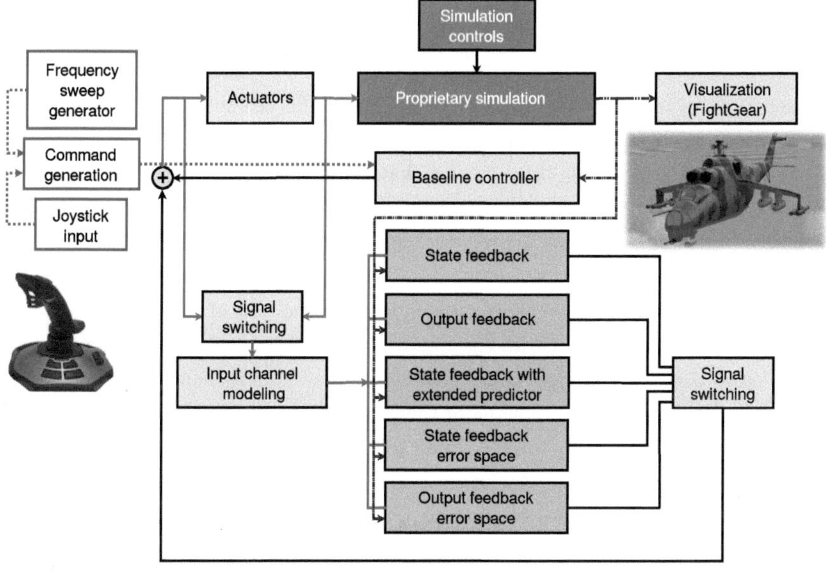

FIGURE O.2: Simulation topology

Figure O.2 shows the simulation architecture.

Hence, flight tests can be carried out without any controller, with the baseline controller alone, with the $\mathcal{L}_1$-controller alone, or with a combination of both.

In aerobatics mode, attitude loops are switched off if some threshold in at least one of pitch or roll angle is exceeded. Reinitialization of the filters and integrators in attitude elements is conducted when reaching normal attitude regimes and low angular rates (otherwise the attitude would move back significantly in reverse direction after fast rates). This is applicable to rate command modes but not to attitude and speed command modes. In addition, in aerobatics mode any attitude protection and turn coordination is disabled.

O.3 Parameter Scheduling in State Feedback

Two options of scheduling of the state predictor parameters a and b are shown in Figure O.3: Linear interpolation or the approximation by analytic functions.

O.4 Actuators

A separation between trim actuators and fast actuators with limited authority is omitted. One set of actuators with full authority is modeled.

A third order model resembles the behavior of hydraulic actuators.

Prior to takeoff, the actuators are driven to somewhere near their trim position of the subsequent hover flight. If a baseline controller *is not* active, this static trim value is added to the actuator input.

If a baseline controller *is* active, this trim value is variable and part of the baseline controller. An attitude angle changing with e.g. airspeed or collective position from look-up tables is added to the attitude angle commanded by the pilot. Additionally, a trim value for the actuators is added to the total actuator command vector, also provided by look-up tables over airspeed and optionally collective lever position.

This approach is convenient in the simulation as the low-cost joystick does not provide trim capabilities. Usually in any real system trim procedures are done differently.

The gain J amounts the percentage of actuator position ($0...100$ %) to degree blade angle; furthermore, $\beta = 2\zeta w_2$ with 'ζ' being the relative damping (E.g. $cmd \, [deg] \cdot \frac{100 \, \%}{20 \, deg}$). With the actuator command being a vector, 'J' can be understood as diagonal matrix with four entries.

This actuator as shown in Figure O.4 models only position and rate saturations. It is assumed that moment saturations are rarely reached. Rate saturations are chosen as $170 \, \%/s$. The bandwidth lies at about $50 \, rad/s$. Furthermore, an actuator delay of $50 \, ms$ is modeled.

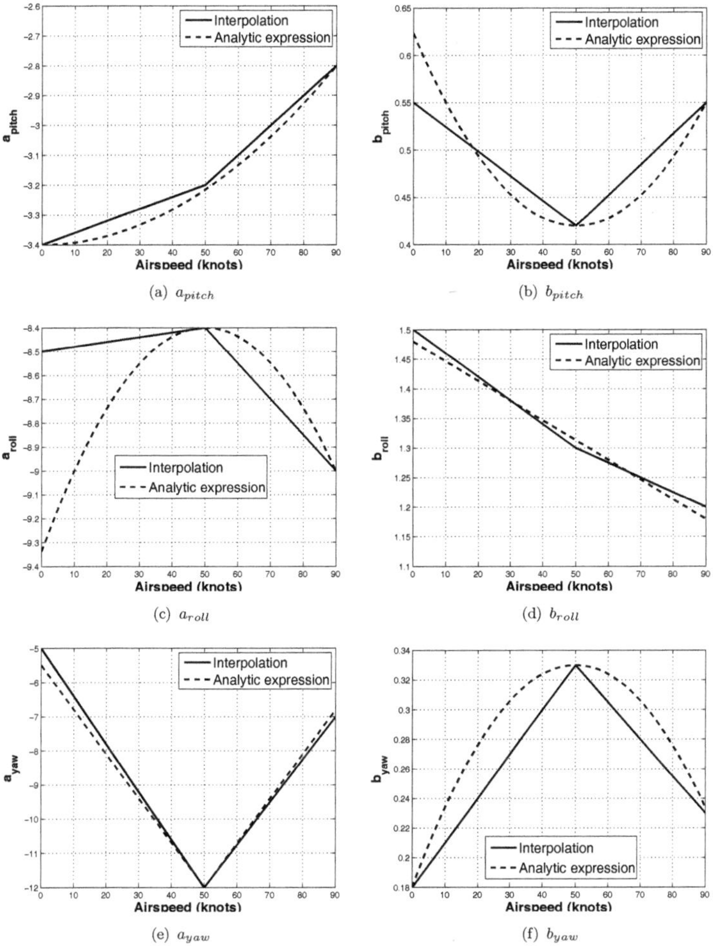

(a) a_{pitch}

(b) b_{pitch}

(c) a_{roll}

(d) b_{roll}

(e) a_{yaw}

(f) b_{yaw}

FIGURE O.3: Parameter scheduling: Interpolation and analytic expressions

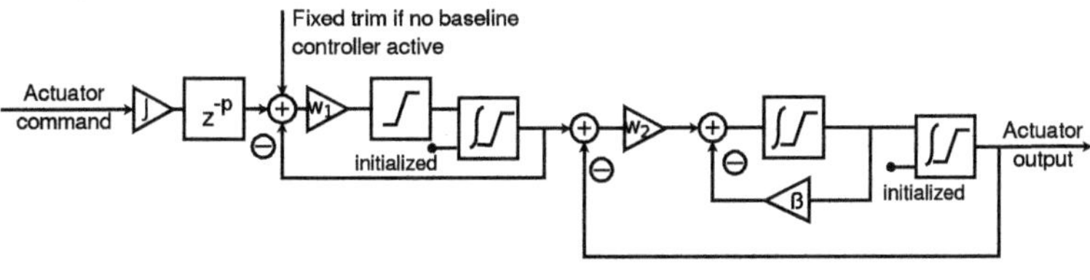

FIGURE O.4: Structure of the modeled actuators

Time delays in the feedback channel from CPU, bus latencies, and mechanical delays are combined into one single modeled delay at the actuator block. The sensor delay is modeled separately.

O.5 Noise Modeling

For emulating the signal quality of real sensors, the simulation output is superimposed by noise. For the case of IMU signals (attitude angles and body-fixed angular rates), two sources of noise are considered:

1) Sensor errors, modeled as randomly distributed noise.

2) Unwanted but correctly measured motions, e.g. vibration, structural effects, etc.

It is expected that the randomly distributed portion due to measurement noise is small compared to vibration levels. In this simulation, an RMS of 0.1 deg/s is applied, which results in a variance of $(0.1/57.3\ rad/s)^2$ implemented.

A vibration signal $\jmath(t)$ is computed as:

$$\jmath(t) = A \cdot sin(n\Omega t)$$

where '$\jmath(t)$' is the amplitude or weighting, 'n' is the order of higher harmonics of the rotor's rotational speed 'Ω'. The following values are used for a four-bladed rotor:

n [-]	2	4	8
$\jmath(t)$ [rad/s]	0.02	0.05	0.03

This means that the average amplitude for modeled vibrations is about ten times higher than the modeled sensor noise. Only the body-fixed angular rates are affected by vibration measurements in this simulation as the effect on attitudes is much smaller due to the high frequencies. The high frequency vibration with $n = 16$ is neglected, as amplitudes tend to decline with higher frequencies and a difference to Gaussian noise would hardly be visible with an IMU sensor rate between 50 Hz and 100 Hz.

O.6 Miscellaneous

Implementing Time Delays:

The time delay 'τ' can be implemented such that it is independent of the sampling time by `ceil(τ/T)`, where 'T' describes the time step length and the function `ceil` (it is called `ceil` in many programming languages) rounds a floating-point number to the next higher integer. The function `ceil(`τ/T`)` delivers an integer number, which is the number of delayed time steps.

System Identification:

System identification is automated. In a command line interface parameters are set. In the simulation, the connection to FlightGear is disabled by default (can be reenabled manually) and with it the "real time module" is disabled for the simulation to run faster. Frequency sweeps are computer-generated. Data of concern are stored.

Command Line Interfaces for System Identification:

The following prompts work with keyboard inputs only to ensure greatest possible compatibility across software and platform versions. Default values are given in "[]". Frequency sweeps can be applied to the rate or attitude command as well as on the actuator input.

Script #1, Generating data:

```
Initial altitude above MSL (m) [100]:
Initial velocity (m/s)  [30]:
Sweep will be applied on signal #  [4]:
theta_ref .. 1    q_ref ...... 4      coll_act ... 7
phi_ref .... 2    p_ref ...... 5      pitch_act .. 8
psi_ref .... 3    r_ref ..... 6       roll_act .. 9       yaw_act  ... 10
Any appendix to the filename?  [ ]:
FOOBAR
Activating sweep generator for command on: RATE ................ [OK]
Initializing .............................................. [OK]
Simulating ...
Simulation complete ....................................... [OK]
Storing Data .............................................. [OK]
```

After loading a data set from the directory of the saved files, a command prompt for system identification is started:

Script #2, Processing data:

```
Axis (pitch(p), roll(r), yaw(y)) [p]:
Minimum frequency to be displayed (Hz) [0.1]:
Maximum frequency to be displayed (Hz) [3]:
Display FDB-plots? [Y/n]:
```

Bibliography

[1] Magnus Bichlmeier, Florian Holzapfel, Enric Xargay, and Naira Hovakimyan. $\mathcal{L}_1$ Adaptive Augmentation of a Helicopter Baseline Controller. In *Proceedings of AIAA Guidance, Navigation and Control Conference*, Boston, MA, USA, 2013.

[2] Magnus Bichlmeier and Florian Holzapfel. An $\mathcal{L}_1$ - Adaptive Vertical Speed Controller for Helicopters in Hover. In *55th Israel Annual Conference on Aerospace Sciences, Tel Aviv, Haifa, Israel*, 2015.

[3] Magnus Bichlmeier. Design and Performance Assessment of a Model Following Control System for Helicopters using Feedback Control Loops. Master's thesis, Technical University of Munich, Munich, May 2011.

[4] Aeronautical Design Standard, Handling Qualities Requirements for Military Rotorcraft. US Army Aviation and Missile Command, USAAM-COM, March 2000. ADS-33E-PRF.

[5] Gareth Padfield. *Helicopter Flight Dynamics*. Blackwell Publishing, Oxford, UK, 2nd edition, 2007.

[6] A. R. S. Bramwell, George Done, and David Balmford. *Bramwell's Helicopter Dynamics*. Butterworth-Heinemann, Oxford, UK, 2nd edition, 2001.

[7] Walter Bittner. *Flugmechanik Der Hubschrauber*. VDI-Buch. Springer, 2006. ISBN 9783540275411.

[8] Anshu Siddarth and John Valasek. Tracking Control for a Non-Minimum Phase Autonomous Helicopter. In *AIAA Guidance, Navigation, and Control Conference*, 2012.

[9] Mark B. Tischler and Robert K. Remple. *Aircraft and Rotorcraft System Identification. Engineering Methods with Flight Test Examples*. AIAA, Reston, Virginia, 2006.

[10] Phillip J. Gold and James B. Dryfoos. Design and Pilot Evaluation of the RAH-66 Comanche Selectable Control Modes. In *AHS Conference on Flying Qualities and Human Factors*, San Francisco, CA, January 1993.

[11] Karl. J. Åström and Richard M. Murray. *Feedback Systems: An Introduction for Scientists and Engineers*. Princeton University Press, 2008.

[12] Gunter Stein. Respect the Unstable. *Control Systems, IEEE*, 23(4):12–25, 2003.

[13] Naira Hovakimyan and Chengyu Cao. $\mathcal{L}_1$ *Adaptive Control Theory*. Society for Industrial and Applied Mathematics, Philadelphia, PA, 2010.

[14] Kumpati S. Narendra and Anuradha M. Annaswamy. *Stable Adaptive Systems*. Information and System Sciences. Prentice Hall, Englewood Cliffs, NJ, 1989.

[15] Evgeny Kharisov, Kwang Ki Kim, Xiaofeng Wang, and Naira Hovakimyan. Limiting Behavior of $\mathcal{L}_1$ Adaptive Controllers. In *AIAA Guidance, Navigation and Control Conference*, Portland, OR, 2011. AIAA-2011-6441.

[16] Justin Vanness, Evgeny Kharisov, and Hovakimyan. $\mathcal{L}_1$ Adaptive Control with Proportional Adaptation Law. In *American Control Conference (ACC), 2012*, pages 1919–1924, 2012.

[17] Chengyu Cao and Naira Hovakimyan. Design and analysis of a novel $\mathcal{L}_1$ adaptive control architecture, Part I: Control signal and asymptotic stability. In *American Control Conference*, pages 3397–3402, Minneapolis, MN, June 2006.

[18] Zhiyuan Li and Naira Hovakimyan. $\mathcal{L}_1$ Adaptive Controller for MIMO System with Unmatched Uncertainties using Modified Piecewise Constant Adaptation Law. In *IEEE Conference on Decision and Control*, Maui, HI, December 2012.

[19] Zhiyuan Li. *Guidance, Control and Estimation of Autonomous Vehicle Systems*. PhD thesis, University of Illinois at Urbana-Champaign, 2013.

[20] Evgeny Kharisov and Naira Hovakimyan. $\mathcal{L}_1$ Adaptive Output Feedback Controller for Minimum Phase Systems. In *American Control Conference (ACC), 2011*, pages 1182–1187, 2011.

[21] Justin Vanness, Evgeny Kharisov, and Naira Hovakimyan. $\mathcal{L}_1$ Adaptive Output-feedback Controller for Linear Time-varying Reference Systems. In *American Control Conference (ACC), 2013*, pages 4183–4188, 2013.

[22] Jie Luo and Chengyu Cao. $\mathcal{L}_1$ Adaptive Output Feedback Controller for a Class of Nonlinear Systems. In *2011 50th IEEE Conference on Decision and Control and European Control Conference (CDC-ECC)*, Orlando, FL, December 12.-15. 2011.

[23] Irene M. Gregory, Enric Xargay, Chengyu Cao, and Naira Hovakimyan. Flight Test of $\mathcal{L}_1$ Adaptive Controller on the NASA AirSTAR Flight Test Vehicle. In *AIAA Guidance, Navigation and Control Conference*, Toronto, Canada, August 2010.

[24] Xin Jin, A. Ray, and R.M. Edwards. Integrated Robust and Resilient Control of Nuclear Power Plants for Operational Safety and High Performance. *Nuclear Science, IEEE Transactions on*, 57(2):807–817, 2010. ISSN 0018-9499.

[25] Xiaofeng Wang, Enric Xargay, Naira Hovakimyan, and Irene M. Gregory. Integration of Dynamic Flight Envelope into the $\mathcal{L}_1$ Adaptive Control Framework. In *AIAA Guidance, Navigation, and Control Conference*, 2012.

[26] Dapeng Li, Naira Hovakimyan, and Chengyu Cao. Positive Invariance Set of $\mathcal{L}_1$ Adaptive Control in the Presence of Input Saturation. In *AIAA Guidance, Navigation and Control Conference*, Toronto, Canada, August 2010.

[27] Vijay V. Patel, Chengyu Cao, Naira Hovakimyan, Kevin A. Wise, and Eugene Lavretsky. $\mathcal{L}_1$ Adaptive Controller for Tailless Unstable Aircraft in the Presence of Unknown Actuator Failures. *International Journal of Control*, 82(4):705–720, 2009.

[28] Jeffrey W. Harding, Scott J. Moody, Geoffrey J. Jeram, M. Hossein Mansur, and Mark B. Tischler. Development of Modern Control Laws for the AH-64D in Hover/Low Speed Flight. In *American Helicopter Society Annual Forum*, Phoenix, AZ, May 2006.

[29] Enric Xargay, Naira Hovakimyan, Vladimir Dobrokhodov, Isaac Kaminer, Chengyu Cao, and Irene M. Gregory. $\mathcal{L}_1$ Adaptive Control in Flight. In John Valasek, editor, *Advances in Intelligent and Autonomous Aerospace Systems*, volume 241 of *Progress in Aeronautics and Astronautics*, pages 129–206. American Institute of Aeronautics and Astronautics, 2012.

[30] Anders Pettersson, Karl J. Åström, Anders Robertsson, and Rolf Johansson. Augmenting $\mathcal{L}_1$ Adaptive Control of Piecewise Constant Type to a Fighter Aircraft. Performance and Robustness Evaluation for Rapid Maneuvering. In *AIAA Guidance, Navigation and Control Conference*, Minneapolis, Minnesota, USA, August 2012.

[31] David Erdos, Tal Shima, Evgeny Kharisov, and Naira Hovakimyan. $\mathcal{L}_1$ adaptive control integrated missile autopilot and guidance. In *Guidance, Navigation, and Control and Co-located Conferences*, pages –. American Institute of Aeronautics and Astronautics, 2012.

[32] Chengyu Cao, Jie Luo, John Cooper, and Irene Gregory. Filter Bandwidth Adaptation in the $\mathcal{L}_1$ Adaptive Control Architecture. 2013.

[33] Evgeny Kharisov. $\mathcal{L}_1$ *Adaptive Output-Feedback Control Architectures*. PhD thesis, University of Illinois at Urbana-Champaign, 2013.

[34] Vittorio Cortellessa, Bojan Cukic, Diego Del Gobbo, Ali Mili, Marcello Napolitano, Mark Shereshevsky, and Harjinder Sandhu. Certifying Adaptive Flight Control Software. In *ISACC 2000: The Software Risk Management Conference*, 2000.

[35] Guido Weber, Tim Lammering, Sven Thierer, Peter Schaedler, Georg Ried, and Tom Schneider. The Liebherr Fully Integrated FCS Design - a Case Study. In *2013 Aviation Technology, Integration, and Operations Conference*. American Institute of Aeronautics and Astronautics, August 2013.

[36] Stephen A. Jacklin. Closing Certification Gaps in Adaptive Flight Control Software. In *AIAA Guidance, Navigation and Control Conference*, Honolulu, HI, August 2008. AIAA-2008-6988.

[37] Naira Hovakimyan, Chengyu Cao, Evgeny Kharisov, Enric Xargay, and Irene M. Gregory. $\mathcal{L}_1$ Adaptive Control for Safety-Critical Systems. *Control Systems, IEEE*, 31(5): 54–104, 2011. ISSN 1066-033X.

[38] Chris Wilkinson, Jonathan Lynch, and Raj Bharadwaj. Final Report - Regulatory Considerations for Adaptive Systems. Technical Report NASA/CR–2013-218010, NASA, 2013.

[39] Ali Mili, Bojan Cukic, Yan Liu, and Rahma Ben Ayed. Towards the Verification and Validation of Online Learning Adaptive Systems. In T. Khoshghoftaar, editor, *Computational Methods in Software Engineering*. Kluwer Scientific Publishing, 2002.

[40] Xiaofeng Wang and Naira Hovakimyan. $\mathcal{L}_1$ Adaptive Controller for Nonlinear Time-Varying Reference Systems. *Systems & Control Letters*, 61(4):455–463, 2012.

[41] John Cooper and Chengyu Cao. $\mathcal{L}_1$ Adaptive Control with Additional Memorizing Mechanism. International Journal of Adaptive Control and Signal Processing, submitted, 2013.

[42] Magnus Bichlmeier. An $\mathcal{L}_1$ Adaptive Output Feedback Controller using Modified Piecewise Constant Adaptation Law. In *Proceedings of AIAA Guidance, Navigation and Control Conference, SciTech 2016, San Diego, CA, USA*, 2016.

[43] Evgeny Kharisov, Naira Hovakimyan, and Karl Åström. Comparison of Several Adaptive Controllers According to Their Robustness Metrics. In *Proceedings of AIAA Guidance, Navigation and Control Conference*, Toronto, Canada, 2010.